Statistical Analysis for Decision Makers in Healthcare

Understanding and Evaluating Critical Information in Changing Times

Second Edition

D0226195

Statistical Analysis for Decision Makers in Healthcare

Understanding and Evaluating Critical Information in Changing Times

Second Edition

Jeffrey C. Bauer

CRC Press is an imprint of the
Taylor & Francis Group, an **informa** business
A PRODUCTIVITY PRESS BOOK

Productivity Press
Taylor & Francis Group
270 Madison Avenue
New York, NY 10016

© 2009 by Taylor & Francis Group
Productivity Press is an imprint of Taylor & Francis Group, an Informa business

No claim to original U.S. Government works

Printed in the United States of America on acid-free paper
10 9 8 7 6 5 4 3 2 1

International Standard Book Number-13: 978-1-4398-0076-8 (Paperback)

This book contains information obtained from authentic and highly regarded sources. Reasonable efforts have been made to publish reliable data and information, but the author and publisher cannot assume responsibility for the validity of all materials or the consequences of their use. The authors and publishers have attempted to trace the copyright holders of all material reproduced in this publication and apologize to copyright holders if permission to publish in this form has not been obtained. If any copyright material has not been acknowledged please write and let us know so we may rectify in any future reprint.

Except as permitted under U.S. Copyright Law, no part of this book may be reprinted, reproduced, transmitted, or utilized in any form by any electronic, mechanical, or other means, now known or hereafter invented, including photocopying, microfilming, and recording, or in any information storage or retrieval system, without written permission from the publishers.

For permission to photocopy or use material electronically from this work, please access www.copyright.com (http://www.copyright.com/) or contact the Copyright Clearance Center, Inc. (CCC), 222 Rosewood Drive, Danvers, MA 01923, 978-750-8400. CCC is a not-for-profit organization that provides licenses and registration for a variety of users. For organizations that have been granted a photocopy license by the CCC, a separate system of payment has been arranged.

Trademark Notice: Product or corporate names may be trademarks or registered trademarks, and are used only for identification and explanation without intent to infringe.

Library of Congress Cataloging-in-Publication Data

Bauer, Jeffrey C.
Statistical analysis for decision makers in healthcare : understanding and evaluating critical information in changing times / Jeffrey C. Bauer. -- 2nd ed.
p. ; cm.
Includes bibliographical references and index.
ISBN 978-1-4398-0076-8 (hardcover : alk. paper)
1. Medical statistics. 2. Statistics. I. Title.
[DNLM: 1. Delivery of Health Care. 2. Statistics as Topic. 3. Decision Making. 4. Models, Statistical. WA 950 B344s 2009]

RA409.B35 2009
610.72--dc22 2009012018

Visit the Taylor & Francis Web site at
http://www.taylorandfrancis.com

and the Productivity Press Web site at
http://www.productivitypress.com

He uses statistics as a drunken man uses lampposts—
for support rather than illumination.

Andrew Lang
Scottish Author
1844–1912

Contents

Introduction

AND NOW, FOR SOMETHING COMPLETELY DIFFERENT ...

You might think I should have my head examined for writing a book about statistics. Why, after all, would any sane person want to spend valuable time adding to the published literature on one of the world's most feared and incomprehensible subjects? Aren't statisticians people who didn't have the personality to become mathematicians? Don't students refer to their classes in the subjects as "sadistics?"

You might also wonder why you should read a book on statistics. If you are a typical healthcare professional, you probably feel you barely survived the required course in statistics; it may even account for the only B⁻ on your transcript. You would like to forget the classroom experience, and you probably have forgotten the subject matter itself. Why take another pass at a subject that offers so little inherent appeal?

The reason for this book is pretty simple. I wrote it and you are reading it because, like it or not, we cannot avoid statistics in a competitive healthcare industry. Every day, we encounter people who want us to change the things we do: update diagnostic equipment, modify employee training programs, add drugs to the formulary, adopt "best" clinical practices, sign contracts to participate in new payment mechanisms, and even reform the entire healthcare system.

The people who would like us to make the changes try to influence us by citing statistics from studies that presumably show why we would be better off if we did things their way. Today's decision makers need to know something about statistics if for no other reason than to deal intelligently with all the studies that suggest changes to something that will make a desired difference in the outcome. If you are never confronted with studies in your day-to-day management of healthcare resources or patients, you probably do not need to understand statistics. But if you do encounter studies in your daily quest to make better decisions, this book is written for you.

Back in the good old days (before the mid-1980s), the executives and clinicians who decided how healthcare would be delivered did not have to pay much attention to studies in general or to statistics in particular. Deciding whether to make a proposed change was a personal philosophical issue more than a practical concern because the marketplace was very generous. Retrospective, cost-based reimbursement and easy cost shifting pretty much allowed hospital managers to do whatever they wanted. Clinicians only had to think about practicing in accord with loosely defined "local standards of care." That old statistics book probably sat on the shelf, unopened since college days.

Today, administrators are managing scarce resources in fiercely competitive markets, and health professionals are practicing in an environment of research-based practice guidelines (because resources are scarce). Those research reports in *Harvard Business Review* or the *New England Journal of Medicine* have taken on a new significance. We now must care whether hiring hospitalists, installing an electronic medical record, or upgrading diagnostic technology really makes a difference because being wrong has consequences. Where can we turn for guidance in making such decisions? To all those published studies, of course. And what do we need to know in order to interpret the studies? Statistics.

I wrote this book because I know that your old "stats" book is not going to help you very much even if you do get around to reviewing it. I have taught statistics and research methods to graduate students in medical, dental, nursing, and business schools for more than 20 years, so I have had semester-long encounters with hundreds of highly intelligent professionals who already had a statistics book from their previous undergraduate statistics class. Most of these students felt comfortable admitting to me that they never really did understand their old statistics book in the first place. They learned to cope with the math of statistics well enough to pass the course, but they never did understand why they were doing what they were doing.

I share my students' frustration because I have faced the difficult task of selecting a text every time I taught the course. I've assigned at least a dozen introductory-level statistics books, and I have never found one that helped the students fully understand the subject using examples from the world of healthcare. My students have convinced me that health professionals need their own statistics book, one that is very different from all the others in both general concept and specific illustrations. Thanks to the constructive feedback I have received from my students, this statistics book is very different, and it is dedicated to them.

KNOWING WHAT AND WHY, NOT HOW

My 35-plus years of experience as a professor and consultant in many different medical settings led me to conclude that today's healthcare decision makers need to have a workable understanding of the theory of statistics—which is something very different from knowing how to perform statistical computations. Consequently, this book *emphasizes basic concepts, not mathematics.* People who are intimidated by equations can read this book without the fear of getting lost in the act of taking square roots of the sum of squared deviations and dividing by the sample size minus one (or is it the sample size minus two?). Knowing which equation to solve and why is vastly more important than knowing how to solve an equation without knowing why.

Computers can do the computations, but only the decision maker can tell the computer which computations to do. (Actually, computers do make computational errors every once in a while, but infinitesimally less often than humans.) Countless students have told me that they spent so much time learning formulas and solving problems in their previous statistics classes that they never had time to figure out what it all really meant. This book explains the "what" and "why" of statistics so that busy professionals can understand and evaluate the statistical claims made in all the studies that come across their desks. Those who still have the time and desire to master computational skills can consult just about any other statistics book, but the extra mathematical exercise will not make them better decision makers.

The computer's clear superiority in solving numerical problems is not the only reason for my decision to stress concept and deemphasize computation in a statistics book for decision makers in healthcare. The practical fact of the matter is that today's managerial and clinical executives are paid to think and to make decisions, not to spend time doing math. Staff employees, research assistants, administrative interns, technologists, and the like are paid to collect and analyze the data. The executive in an information-rich society, appropriately identified by Harlan Cleveland as the "get-it-all-together person," must know enough to evaluate the appropriateness and the quality of their work:

> The get-it-all-together person needs above all to be good at judging whether the experts who stream through the executive office, creating a *chronic condition of information entropy on the executive's desk* [italics mine], are functioning as competent experts.[1]

I know that my decision to write a book that explains the basic theory and largely ignores the math will offend many statisticians who have spent years pursuing computational elegance in their field. Unfortunately, their intellectual purity is not very helpful to health professionals who only need a practical understanding of statistics. Most of us need nothing more than the statistical equivalent of a Berlitz phrase book and a Rick Steves travel guide so that we can get along and not make stupid mistakes in a foreign land. By comparison, the typical statistics book is the equivalent of a scholarly treatise on linguistics and cultural anthropology.

CHALLENGING THE FAITH

In my experience, courses and textbooks in statistics are generally dedicated to the proposition that the student must believe, but not necessarily understand, statistics. This book endeavors to do the opposite: to make statistics understandable but not necessarily believable. If this book is successful, statistics will lose its element of mystery. You will no longer be afraid of statistics. Some—possibly much—of your previous awe for statistics will be replaced by cynicism. You will have a strong defense against the many people who try to intimidate you with statistics because you will know the weak points in their offense. You will also develop considerable respect for those who use statistics properly.

---

[1] Harlan Cleveland, *The Knowledge Executive: Leadership in an Information Society* (New York: Dutton, 1985), p. 11. This is one of only two books I have used year after year in my statistics classes; I heartily recommend it. Regrettably, it is not a recent publication, so look for it online or in used bookstores if necessary.

My goal is based on an analogy that has often come to my mind, one of native heathens (statistics students) being converted to a new religion by foreign missionaries (statistics teachers). In the traditional approach, the converts accept the beliefs, even though few will ever understand the historical or cultural context of their adopted religion. In contrast, my approach is like that of a college class on a religion other than your own. I want you to understand the catechism of the other religion (statistics), but I do not expect you to convert to it just because you get an A in the course. I hope you will understand statistics enough to recognize both its usefulness and its limitations. You should emerge from the experience enriched but skeptical, because statistics is not based on any demonstrable truth that God gave to only one highly educated caste of his or her people.[2]

Of course, students who are training for careers in the health professions tend to be intelligent. They can learn to solve statistical problems once they see that the underlying math requires nothing more than the computational skills they learned in high school—even though they may struggle for a while as they decide which formula to use in a particular situation. To my dismay, I have noticed that very few students ever move beyond computational competence to understand what the formulas actually represent. They believe implicitly that the equations express wisdom only accessible to the priesthood with a Ph.D. Since these students are taking statistics because it is required to get a degree in some other field (such as health or business administration, public health), they are happy to believe that statistics is built on a solid conceptual base that they do not have to understand.

The theory of statistics is built on a few key assumptions that are explored in-depth later in this book. The real-world power of statistical analysis cannot be any better than the real-world relevance of the underlying assumptions. Consequently, understanding the assumptions behind statistical theory is the first critical step toward the competent use of statistics. Stated in another way, being able to compute a statistical test cor-

---

[2] Many statisticians will undoubtedly label me a heretic for explicitly pursuing such a disrespectful goal. Heresy is actually becoming quite respectable in the mathematics-based fields of science. As evidence, one needs look no further than to the current collection of books on cosmology and origins of the universe. The once solid worldview of Albert Einstein has been challenged over the past few years by many living Nobel laureates who offer very different concepts of physical reality (including the view that reality is not what we think it is). Physics has become a less-certain science as the tools of mathematics and data analysis have improved dramatically.

rectly is irrelevant if the assumptions behind the formula are inconsistent with the situation to which the statistic is being applied.

The problem is rooted in the sad fact that books on introductory statistics seldom explain the underlying assumptions or their importance. Even worse, those books that do allude to the link between theory and practice often say that the theory can be ignored because it is either too restrictive or does not matter very much! The result is a lot of bad studies done by researchers with just enough statistical knowledge (or too little professional integrity) to be dangerous. My central purpose in writing this book, then, is to teach you to separate the good from the not-good. Sad to say, you will find that good statistical practice is not the norm. At least you will know how not to be fooled, which can save you from making decisions based on bad statistics.

STRUCTURE AND CONTENT OF THIS BOOK

The modern "science" of mathematical statistics began to develop in the mid-1700s to meet the needs of a growing number of people who were making quantitative observations—that is, collecting data to advance their understanding of the world around them. Bureaucrats and scientists started counting things like never before, which created a need to develop methods for summarizing large collections of numbers.

Section I: The Scientific Foundations of Statistical Analysis

The historical relationship between science in general and statistics in particular is fundamentally important to the understanding of either, so this book begins with an exploration of the scientific foundations of statistical analysis. Chapter 1 reviews the universal principles of the scientific method because they provide the necessary but commonly ignored foundation of good statistical practice. Chapter 2 presents a careful review of experimental design because a proper experiment should lie behind any effort to influence a health professional's managerial or clinical decisions on the grounds that something makes a difference. Together, these two chapters provide the reader with an essential first line of defense against the many studies that are not built on solid foundations.

SLIDE-RULE SCIENCE IN THE COMPUTER AGE?

Very few people ever become aware of the fact that statistics' fundamental assumptions date from the eighteenth and nineteenth centuries (that is, the 1700s and 1800s). Almost alone among scientific disciplines, statistics changed very little during the twentieth century. However, most fields of science are now being redefined every 5 to 10 years, if not faster! Computers have sped up mathematical computations and enabled researchers to analyze larger and larger databases—now doing analyses in seconds, which took hours or days only a few years ago. Yet, the underlying theoretical foundations of statistics have been pretty much the same for more than 100 years.

Curiously, statistics in the age of computers is still conceptually confined to computations that can be performed on a slide rule. A slide rule, for younger readers who grew up in the digital age, was a mechanical device that resembled a ruler. It used logarithmic scales for multiplication, division, and exponential computations.

Nerds like me who went to high school and college in the 1960s or earlier had to use mechanical (that is, big, slow, and noisy) adding machines or pencil and paper to do the addition and subtraction in statistical computations. The slide rule handled the squares and square roots. The process was slow and far from error free, so we thought twice before performing a statistical analysis.

Sadly, computers and hand-held digital devices have made statistics too easy to do now. Problems that took hours to set up and solve just a few decades ago can now be handled in nanoseconds by people who do not have the slightest idea what they are doing because the computer requires them to do nothing more than enter some data. By virtually eliminating human involvement in doing statistics, computer power has done us a disservice.

Happily, the digital revolution has a very positive dimension. With its multimedia interfaces, the modern computer now makes possible many new ways of looking—both literally and figuratively—at numbers and other forms of measurement. Data can also be analyzed over time, revealing dynamic effects that are obscured or completely lost in the static realm of statistics. The computer offers so many new approaches to quantitative and qualitative analyses that traditional statistics may soon be obsolete, particularly in genetics and molecular medicine.

Section II: The Fundamental Importance of Data

Good science is an essential precondition of good statistics, but it is not solely sufficient to justify putting faith in the results of a statistical study. The data must also be good. Since bad data are even more common than bad science, this section spends considerable time helping you decide whether to have confidence in the numbers that are inserted into statistical formulas. Chapter 3 is devoted to a general discussion of the quality and types of data. Chapter 4 takes a special qualitative look at sampling and surveys since so many of today's studies are based on data obtained via survey research. Together, the two chapters in this section will give you one of your most powerful weapons in the never-ending fight against statistical malpractice.

Section III: The Different Types of Statistics

Much to my amazement, I discovered that most students taking statistics for the second or third time have not already learned that there are different categories of statistics to accomplish different analytical purposes, just as there are different categories of drugs to accomplish different biochemical outcomes (such as antibiotics to kill bacteria, psychotropic medications to alter brain function, thrombolytics to dissolve clots, and so on). Happily, I discovered that statistics starts to make sense when students are exposed separately to each of the analytical branches of statistics; therefore, the four chapters in this section explain the models for describing populations and samples (Chapter 5: "Descriptive Statistics"); making inferences about the likelihood that samples are from different populations (Chapter 6: "Inferential Statistics"); studying the relationships between measurements (Chapter 7: "Relational Statistics"); and using models to explain and predict relationships (Chapter 8: "Explanatory Statistics").

And now, for something completely different ...[3]

---

[3] Although this phrase is not protected by copyright, I do acknowledge Monty Python's John Cleese for popularizing it. I am a real fan of the work of this British comedy troupe, and some of its irreverence is no doubt reflected in my worldview. Statistics and statisticians would be a great subject for a Monty Python skit if the group were ever to get together again.

Section I

The Scientific Foundations of Statistical Analysis

The mystery that surrounds the world is eternal. (*Le mystère qui envelope l'univers est éternel.*)

Louis Pasteur
Speech to l'Académie Française
Oeuvres, Tome VII (Masson & Cie., 1939)

1

Scientific Method: The Language of Statistical Studies

Good science and good statistics must go hand in hand. As the old song goes, you cannot have one without the other. Unfortunately, many researchers and policy analysts who try to influence our decisions with their studies have not learned to respect the fundamental importance of good science. (Actually, I suspect that quite a few studies are presented by people who do understand the importance of science but choose to ignore it.) Studies built on a flimsy scientific base are remarkably common, to the point of being considered normal. Healthcare decision makers must develop the knowledge and the courage not to take seriously the many studies that fail to pass muster on scientific grounds.

Very few statistics books pay any attention to the fundamental issues of scientific integrity in the data-collection process. Most authors of statistics books apparently assume that the data to be analyzed are collected in a scientifically defensible manner, so their books move immediately to the first steps in the process of organizing data for analysis.[1] This shortcut has allowed unsophisticated or unscrupulous researchers to conduct and publish studies without giving a second thought to the scientific conditions that should be met before any data are collected, analyzed, and reported. *Good statis-*

[1] I happened to be teaching at the School of Medicine at the University of Wisconsin–Madison when I originally wrote this sentence, so I went to the university bookstore on campus to test my hypothesis. Upon examining approximately 15 statistics books stocked in the sections for Statistics and Mathematics, I found only two that explicitly included chapters on the importance of the scientific integrity of methods used to collect data. I was pleased to find many good books on research issues in the General Reference section, but I saw no evidence that these books were being used in the introductory courses offered to students who are majoring in something other than statistics.

tics cannot compensate for bad science, so this book begins with a careful examination of the essential scientific foundations of statistical analysis.

As a healthcare decision maker who does not relish the consequences of being misled by bad information, you must care about the scientific base of any study you review. Your first line of defense should be an assessment of the scientific integrity of all studies that are intended to influence your managerial or clinical actions. Always evaluate a study scientifically first, before deciding whether to be impressed with its claimed level of statistical significance or explanatory power. If the study does not conform with the basic principles of good science, you do not need to pay any attention to the study's statistics because—I will say it once again—good statistics cannot compensate for bad science.

Does bad science happen often enough that today's healthcare decision makers need to be concerned about it? Yes, sad to say; the problem is widespread. I find major violations of the scientific method in the majority of statistical studies that I review on almost a daily basis as a medical economist, health futurist, consultant, and expert witness.[2] And this is the same literature aimed at healthcare decision makers! Many researchers are so focused on analyzing and presenting their data that they forget to make sure the data were collected in a scientifically defensible manner.

The problem is exacerbated by the fact that journal editors do not always do a good job of ensuring the scientific integrity of what gets published. Both researchers and editors should be concerned about deficiencies in the scientific foundations of published reports if for no other reason than they would undoubtedly be quite offended to have their studies labeled "unscientific." However, lots and lots of studies are conducted in a most unscientific manner.

[2] Conducting a careful literature review in the late 1970s sensitized me to the questionable scientific base of studies intended to influence the allocation of healthcare resources. In reviewing approximately 40 published studies of the demand for dental care, two colleagues and I found the results of most of the studies were cast in serious doubt by deficiencies in scientific method. See Jeffrey C. Bauer, Arthur P. Pierson, and Donald R. House, Factors Which Affect the Utilization of Dental Services (Hyattsville, MD; U.S. Department of Health, Education, and Welfare, Public Health Service, Health Resources Administration, Bureau of Health Manpower, Division of Dentistry, 1978; Publication No. HRA78-64). I continue to review research reports with the same critical eye, and I do not perceive that the overall situation has gotten any better in the intervening years. Reports of clinical research have improved perceptibly over the past decade in quite a few major journals, but studies of policy and management issues are generally no better now than they were in the 1970s.

How can bad science happen? How can so many people conduct statistical studies using methods that do not adhere to the basic principles of science? I think that the common lack of attention to the scientific integrity of studies derives from the fact that statistics is a required course, while a course in research methods is not. Formal courses in the proper methods of scientific research are exceedingly rare at either the undergraduate or graduate levels. Students are not forcibly exposed to the fundamentals of good science.[3]

This sad state of academic affairs makes about as much sense as allowing graduate students in comparative literature to write about the differences between English and French novels without expecting them to learn French. Too much meaning is lost in translation, especially if the translator is not fluent in the original language. You have to understand grammar and syntax and complex contemporary usage (for example, slang) in two languages in order to translate well from one to the other. Too many people approach statistical studies at the level of "translators" who do nothing more than look up words in a foreign language dictionary. So let us start our inquiry into interpreting studies by becoming fluent in its original language: science.

SCIENCE AS A UNIVERSAL LANGUAGE

Figuratively speaking, the metaphorical description of science as the common language of research is appropriate and useful. Indeed, science is often described as a universal language because its principles are the same throughout the world, and science seeks to understand and explain the way the world works. Discoveries that are made in adherence to its universal principles are immensely helpful to people who make decisions in all fields, including healthcare.

More to the point of this book, the field of statistics was developed over the past few centuries by people with mathematical skills to meet scientists' and public officials' needs to analyze the data they collected through

[3] Indeed, I learned research methods by virtue of being a research assistant with an international group of atmospheric physicists for several years before I became a health economist. My own, very typical Ph.D. training in the social sciences did not include a single course that formally addressed the scientific preconditions for statistical analysis.

experiments and government activities (for example, collecting taxes or conducting the census). Statistics, in other words, is a tool for doing science, both public and private. It is not an alternative to science in the sense that we can study a problem using either science or statistics. *Data should be collected scientifically and analyzed statistically.*

Like all nerds who attended high school in the early 1960s—at a time when science was stressed like never before (and, sadly, never since) because our government was obsessed with beating the Soviets in the race

NEGATIVE KNOWLEDGE

Science is respected worldwide as the civilized method for solving problems, and we rightly expect much of it. The "clockwork" worldview of famous scientists from Sir Isaac Newton to Albert Einstein generally leads people to believe that science exists to reveal the secrets of a logical, orderly world.

However, we must realize that some problems have no solutions; some questions have no answers. A growing number of scientists now hold a "quantum" worldview, one where disorder and randomness are believed to be common. Science cannot find the answer to a question or the solution to a problem in a chaotic system.

I think that chaos is now rather common in our medical system. To paraphrase the title of one of my other books, *Not What the Doctor Ordered*, the health system is no longer organized solely by doctors. Many of the thorny problems now confronting managers and clinicians come from the fact that no one is in absolute control any more. If the ongoing debates over healthcare reform prove anything, it is that the American healthcare system is not structured around any coherent logic or inherent order. (Indeed, many commentators argue that it is not a system at all.)

Decision makers need to be smart enough to recognize unsolvable problems or unanswerable questions. Be wary of studies that claim to have found predictable order when you have reason to expect the opposite. Learn to value negative knowledge: Know what not to do and when not to do it. In such instances, science cannot save us. As some sage once said, anything not worth doing is not worth doing well.

to put a man on the moon—I was taught in numerous courses that science was a universal language because it created a common bond in communities of open-minded inquiry around the world. Much to my surprise, this lofty sounding claim turned out to be true.

I worked for seven consecutive summers (1963–1969) as a research assistant on a large-scale scientific study of hailstorms. The principal scientists came from all over the world. They had to use several different languages for communicating with one another, but whether I was working on an experiment with an Italian or an Australian or a Russian or a French Canadian, I discovered that all the researchers shared the same definitions of experimentation, measurement, controls, and other fundamental concepts of science. Any deviation from these shared principles would have been considered a serious breach of etiquette. Indeed, adhering to the principles of science was probably the only area of complete agreement in this group of international scientists!

We should expect the same loyalty to universal scientific principles in our own country and in our own health professions. Whether healthcare researchers are from Massachusetts or Nevada, from a medical school or an insurance company, from a university or a think tank, or from internal medicine or orthopedic surgery, all ought to be held accountable to the same concepts of science. Science does not have the equivalent of regional dialects.

In similar fashion, scientific method should not vary with researchers' professional backgrounds. No author of a study should be allowed to get away with scientific nonsense on the grounds that he or she was trained to look at problems in a different (that is, nonscientific) way. A study performed by a health economist with a Sc.D. from MIT should be judged according to the same scientific principles as a study conducted by a physician with an M.P.H. from Duke or a nurse with a Ph.D. from the University of California.

The common principles of good research have stood the test of time, so let us begin the task of evaluating statistical studies by discussing the common foundation of all research: scientific method. We will then look at the proper contents of a scientific report. Both these sections provide the information you need in order to decide whether to pay attention to any given study. If a study does not pass muster according to the criteria presented in the following two sections, you do not need to take it seriously—indeed, you probably should *not* take it seriously—because (one last time, I promise) good statistics cannot compensate for bad science.

THE ATTRIBUTES OF SCIENTIFIC METHOD

The scientific method has been defined and analyzed extensively.[4] Although different authors may describe scientific method in slightly different terms, the basic concepts can be summarized under the five common concepts that are the headings of the following discussion. The order of presentation does not matter because all are important (that is, the following attributes of good science are not presented in priority order). A violation of any single element in the scientific method can be enough to make an entire study unworthy of your attention. Like a chain, science is no stronger than its weakest link.

Science Is Open-Minded

Good scientists approach every study with a completely open mind, which means two things. First, scientists are prepared to change their minds based on new information.[5] Second, they are prepared to see the unexpected when they conduct an investigation. Scientists approach their work with a sense of wonder, knowing that they may miss something important if they are only willing to see what they expect to see. For example, Sir Arthur Fleming's famous mistake is a well-known example of the importance of open-minded science. What he found in the Petri dish was theoretically impossible; it just could not be. If he had not taken the time to try to understand it, we might not have penicillin.

[4] I have for many years recommended two well-known books for students who want to learn more about the structure and meaning of science. My favorite general introduction to the subject is W. I. B. Beveridge, *The Art of Scientific Investigation* (New York: Vintage Books, 1950). The philosophical foundations of science are explored in considerably more depth in Thomas S. Kuhn, *The Structure of Scientific Revolutions* (Chicago: University of Chicago Press, 1962), which is the primary source of much of the popularized thinking about paradigms and paradigm shifts.

[5] Rudolph Fleisch's *The Art of Clear Thinking* (New York: Harper & Row, 1945) is a concise classic that deserves renewed attention because it shows how to apply scientific principles to our everyday lives.

Many of our most important scientific discoveries were totally unexpected. Consequently, a meaningful study does not start out with a preconceived notion. (As we will see in the next chapter, a meaningful study does start out with a hypothesis stated in such a way that it does not influence observational judgment.) Therefore, become immediately suspicious of any study that includes statements such as these:

- *It is known that...*
- *We set out to prove that...*
- *We deleted the inconsistent data...*

Finding these phrases and others like them in well-known journals is not uncommon, so be vigilant. The more closed the mind of the author, the less valuable the study. A good researcher expects the unexpected. Open-mindedness should be apparent in all aspects of a study.

Science Is Free of Values

We all know the problems associated with discussing religion and politics. Our parents warned us that these are not safe subjects for polite conversation because many people are unwilling to accept the validity of contrary positions. Discussions of religious or political issues can get in the way of friendships or necessary day-to-day interactions, such as getting along with fellow workers.

Religion and politics can also get in the way of scientific research. For example, scientific research has consistently shown that the AIDS virus (HIV) is transmitted through sexual contact. Stopping AIDS will require more knowledge about human sexual behavior among people in at-risk populations. Yet some people want to distort or prevent research in this area because it clashes with their values. We cannot properly study the problem when research is influenced by the power of individuals whose political or religious values on sensitive subjects such as sexual behavior get in the way of open-minded inquiry.

Science needs to be free to pursue truth through the use of proper methods. Good research cannot be conducted when values lead to a distortion of the scientific method. Jacob Bronowski, one of the last century's most respected scientists, stated that science "is not a set of findings but a search for them. Those who think that science is ethically neutral confuse the findings of science, which are, with the activity of science, which is not."[6] In other words, good scientists do not color their observations with ethical precepts, but personal convictions may determine the areas of their research.

Most of the scientists I know chose their profession because they have strong personal values such as a commitment to saving the Earth or improving human health. The way in which they conduct the research, on the other hand, is not influenced by their values. Indeed, a study should be conducted in exactly the same way by a politically liberal agnostic or a politically conservative Christian.

Healthcare decision makers should be wary about relying on the results of research that was possibly or obviously influenced by nonscientific values. In particular, always pay close attention to the organization or individual who financed the study. When research is funded by an organization that stands to gain or lose from the findings, the sponsor's values are likely to have influenced the study or the way it was reported. Be particularly wary if the source of funds is not identified at all. A good research report always tells you who paid for the study. Failure to identify the source of funds should immediately make you suspicious about the reported results.

Science Is Thoughtful

Good scientists are good thinkers. They spend a lot of time thinking before they conduct experiments. In fact, some of the world's most respected scientists (such as theoretical physicists, molecular biologists) spend all their time thinking, leaving the actual research to someone else. Good scientists think about the problem, the research method, the collection of data, the analysis of results, and even the usefulness of the study. They concoct dozens of different experiments in their minds before they decide which one to conduct in their laboratories. More than one philosopher or historian of science has argued that the scientist's mind is his or her most important research tool.

In addition to recognizing the importance of pure thinking as part of the art of solving problems, good scientists also have the ability to synthesize information—to merge thoughts and concepts from diverse sources into a coherent whole. They are commonly multidisciplinary. The world's

[6] J. Bronowski, *Science and Human Values* (New York: Harper & Row, 1965), pp. 63–64.

greatest scientists tend to be well-rounded individuals. Their ability to think clearly and usefully about the world around them is aided by a working knowledge of art, literature, music, and even sports.[7]

Sadly, today's published literature in medical and health-policy journals frequently lacks evidence of careful, multidisciplinary thought. Studies that pile up on the desks of healthcare decision makers are often ill-conceived and narrow-minded from the beginning because many researchers are in a hurry to get their findings into print and to advance within their field. Pressure to gain recognition through publications and intense competition for limited funding have caused quantity to become more important than quality. This environment creates overspecialization, to the detriment of common sense. (I have always been amused by the definition of a specialist as a person who keeps learning more and more about less and less until he or she knows absolutely everything about nothing.)

To avoid falling under the influence of a thoughtless study, look carefully for signs that the authors regularly spent time thinking while conducting the study. Doing this will require you to spend time thinking about the study too, so make thinking a part of your personal process of evaluating a study that might matter to you. Use common sense. Draw upon your knowledge in a variety of fields. Ask yourself the obvious questions about the work.[8] If plausible answers are not provided in the study, you have probably caught the authors with their thoughts down. The more a study is lacking in answers to questions that are obvious to you, the less you should allow it to influence your managerial or clinical decisions.

Science Is Reproducible

Good scientists are skeptical by nature. They require a lot of evidence before they consider anything a proven fact. Being open-minded, they also

---

7 If you would like to learn more about the important link between good science and broad knowledge, I recommend Arthur Koestler, *Janus* (New York: Vintage Books, 1979). Silvano Arieti's *Creativity: The Magic Synthesis* (New York: Basic Books, 1976) is another interesting and accessible classic work on the mental processes of scientific thinking.

8 If you could use a little help in preparing for the task of asking commonsense questions about studies, you absolutely owe it to yourself to read Darrell Huff's *How to Lie with Statistics* (New York: W. W. Norton, 1954). This short, well-written, and cleverly illustrated paperback is a classic in every sense of the word. It was the other required book in my statistics courses, along with Harlan Cleveland's *The Knowledge Executive: Leadership in an Information Society* (New York: E. P. Dutton, 1985).

accept the possibility that a fact may be disproved at a later time.[9] A collection of consistent facts can add up to a well-established theory. Before it can be elevated from theory to law, a particular phenomenon must be demonstrated over and over again by independent researchers.

The concept of reproducibility (also known as replication) is relevant and important at all stages of the development of science-based knowledge. Good scientists would never accept as "fact" something that had been reported in only one or two studies. The same or very similar findings need to be reported in several corroborative studies before a conclusion becomes widely accepted within the scientific community.

Credibility comes with replication. In other words, one study by itself proves absolutely nothing to a scientist. Just remember the furor created a few years ago over a published report about cold fusion. The discovery of thermonuclear reaction at "room temperature" would have been one of the biggest scientific events of the century if it had been true, but no one else could get the same results when they repeated the experiment. The claim quickly lost its front-page status as scientists returned to their laboratories to rethink the possibilities, and the report was retracted. (Retractions of published studies occasionally occur when their findings cannot be reproduced by other researchers.)

People in the healthcare business tend to assign far too much significance to the findings of a single study. Healthcare's equivalents of cold fusion tend to hang around for a long time. National health policy, concepts of health, or the demand for medical services can be shifted almost overnight by the latest research article published in the *Journal of the American Medical Association* (*JAMA*), the *New England Journal of Medicine*, or *Health Affairs*. Clinicians and managers who want their decisions to be influenced by good science must learn not to jump to conclusions until the effect of a proposed change has been confirmed by several studies.

I suggest a few rules of thumb to help you assess articles with respect to the scientific criterion of reproducibility:

[9] Indeed, with ongoing advances in technologies to make observations in unprecedented detail and to analyze data with remarkable speed, today's scientists even expect facts to be short-lived. The conceptual foundations of most health sciences, for example, are being constantly revised as research scientists learn how to use the research tools that were developed for the Human Genome Project.

- *Do not base your decisions on a single study.* Insist on seeing several consistent reports before you start to think of something as a fact that ought to influence your decisions.
- *Look for a preponderance of evidence.* Half a dozen reports with the same findings are a lot more powerful if they are not contradicted by an equal number of reports with opposite findings. Be aware of the extent of evidence to the contrary, and act appropriately.
- *Expect independent confirmations.* Be extremely cautious in the all-too-common situation where all six corroborating studies were conducted by the same researchers or the same institution.
- *Do not rely on secondhand summaries of studies.* Newspaper and television reports of the latest studies tend to be pathologically simplistic. If a study is important to you, read it yourself in its original form.

Science Is Honest

Since science is a quest for the truth ("the body of real things," according to *Webster's*), its practitioners are expected to be honest. Fudging data or otherwise falsifying observations is simply not acceptable practice in science. The principle of reproducibility will reveal corrupt practices sooner or later, but dishonesty nevertheless occurs in research. It occurs often enough that suspicion is in order when something about a study just does not add up.

Acts of scientific dishonesty have been widely reported in the press over the past decade, and a few elected officials have politicized the issue quite effectively. Scientists' reputations have been tarnished as a result of all the attention, seriously damaging the credibility of work in a few fields. Many prominent research institutions have subsequently developed strict guidelines to distinguish between acceptable and unacceptable ways of conducting a study since some researchers accused of dishonesty have defended themselves by arguing that their suspect actions were common practice.[10]

[10] The controversy over honesty in science and steps to address it are covered extensively in *Science*, an internationally respected journal published weekly by the American Association for the Advancement of Science (AAAS). Reading *Science* and belonging to AAAS are worthwhile for anyone interested in general issues of research and related public policy. For information, visit www.aaas.org.

Like anyone relying on the results of scientific research, the healthcare decision maker cannot easily check the integrity of the researcher who conducted a study of particular interest. The benefit of doubt is in order, and the situation will likely improve as the scientific community gets even more serious about policing itself. However, the possibility of dishonesty should not be ignored. When your suspicions are aroused, you should pay extra attention to the issue of reproducibility to see if the results are supported by the findings of independent researchers.

In extreme situations where a suspicious study is really important to your decision making, you may want to ask the researchers for a firsthand opportunity to review the data so you can judge the work for yourself. Also, be wary of anyone who conducts research in secret when secrecy is not necessary to protect proprietary information or patient confidentiality.

Remember that science is a universal language with common characteristics that should be identifiable in any research that might cause you to change the way your run a medical organization or treat a patient. If a study does not pass scientific muster, you do not need to spend time evaluating the data because the study is flawed in ways that cannot be corrected by statistical analysis. Science comes first, statistics second. Learn to evaluate studies in that order so you will not commit the errors of people who look only at the statistics—people who fail to realize that the study might be built on a foundation of sand.

THE ATTRIBUTES OF A SCIENTIFIC REPORT

Good scientific journals around the world lay out their research articles in an identical or remarkably similar manner that has been around for many, many years. The resulting consistency is extremely useful for a variety of reasons ranging from reader convenience to researcher integrity. For example, skimming an article is much easier and faster when the article adheres to the standardized format discussed in the following pages. (Taking a quick first look at articles is absolutely necessary in a world of so much literature and so little time. Thank heavens for standardization that allows us to decide quickly which articles merit full and careful reading!) A researcher who is more interested in the methodology than in the results of a particular study can immediately find the relevant information because it is clearly identified and located in the expected place. In the absence of a standard format, interested individuals would have to read an entire article just to find the one or two sections of immediate interest to them.

More importantly, the uniform method for reporting scientific studies serves as a checklist that enhances the chances that the inquiry was conducted in accord with the principles of science. The sections of the standard reporting format are roughly equivalent to the basic steps in science, so a researcher is less likely to ignore an important aspect of good research while conducting the study if she or he knows that the written report should address that step.

Failure to follow the standardized format of scientific writing is a warning signal suggesting the possibility, but not the certainty, of significant flaws in a study. An unconventional report can be a sign of problems ranging from the author's poor training or the editor's sloppiness to a conscious effort by either to divert attention from a serious deficiency in the research. Therefore, part of your evaluation of a study needs to include an assessment of the reporting format. You will be able to do this very quickly with a little experience.

Do not hesitate to consider the possible implications of deviations from common practice as you compare a report's structure with the standard format presented below. However, expect to find some variation in the subject headings that provide a report's structure. Most journals use organizational structures similar to the generalized version presented here, but the specific wording and order may vary a bit. The important thing is making sure that all the topics are addressed.

You must also realize that adherence to the standard format is not enough to guarantee that a study is scientifically sound. It is a good sign but not a conclusive one. A report can have all the right parts, but the parts themselves can be flawed. Therefore, examining each of the parts in appropriate detail is also an important step in deciding whether a study is good enough to influence the decisions of a healthcare manager, clinician, or policymaker.

Abstract

A good research report begins with an abstract: a clear and concise summary of what was done and what was learned. Reading the abstract should give you enough information to decide whether you want to read the entire article. I particularly like the abstract formats used by the *Journal of the American Medical Association* and the *New England Journal of Medicine*. Consult a recent issue of either journal to review their formats if you are not already familiar with them. Concise and useful summaries of key articles are now included at the beginning of many scientific journals, but they do not include enough information to make qualitative assessments.

Introduction and Problem Statement

The opening section of the article's narrative text should provide a clear description of the general subject area being addressed by the study. In the first several paragraphs, the authors ought to provide a context for the work by describing the underlying real-world problem and the difference that could be made by understanding it. (If a study is not intended to produce information that might make a difference, why bother with it? Pure research inquiry conducted solely for the sake of knowledge without any expectation of practical application is important in many fields of science, but it is very uncommon in the areas of interest to healthcare decision makers. This book addresses the realm of applied research: inquiry aimed at solving problems.)

An identifiable problem statement is a particularly important component of a good study, so look for it as you begin your assessment of a written report. If you cannot figure out exactly what the researchers were trying to do or why they were doing it, the odds are pretty good they did not know either. A study without a problem statement is like an answer without a question.

Also, watch for studies that begin with a clear problem statement and then go off in a different direction to find a solution that has little or no relationship to the problems as originally stated. I frequently find this "bait and switch" practice in published studies on major issues of health policy, so be prepared for it. For reasons to be discussed in the next chapter (for example, the importance of an *a priori* statement of the research hypothesis), a conceptual link between the problem and the solution is just as important as a careful statement of the problem.

Either flaw—an answer without a question or an answer to a different question from the one asked in the beginning—happens often enough that you need to be attentive to the possibility when you are evaluating studies that might influence the way you solve your administrative or clinical problems. The only studies that are truly useful to you are those that relate directly to your problems and present findings consistent with those problems. This is not necessarily the same thing as "finding solutions to those problems." A study can be very useful by providing information about a wrong solution to a problem that is important to you. Never underestimate the potential value of negative knowledge, knowing what not do!

Review of Literature

Except in the occasional instance where an article is reporting the very first study in an area, it should begin with a review of the published literature on the subject. The review should summarize the findings of other researchers who have investigated the same or similar issues in enough detail to give you a sense of current thinking in the area. The review should be clear and concise.

The references should be footnoted with complete citations (increasingly, URLs and other online links) at the end of the article so you can retrieve and review the previous studies for yourself if you so desire. Paucity or absence of citations is generally a sign of shoddy work. A study that includes little or no review of the literature is suspect when other research on the issue has previously been published.

A study's review of literature should be balanced, including references to key articles on all sides of the issue when contradictory findings have been reported. Mentioning alternative views is part of being open-minded and honest. This goal is often accomplished quite usefully in review articles, so be pleased when you find references to reviews that discuss the published literature in depth. Be cautious when a study cites only articles that are consistent with its findings if other conclusions are known to exist, and be even more cautious when most or all of the citations are previous studies done by the same authors of the study you are reviewing.

In addition to being balanced, a good review of the literature also includes references to studies published in a variety of journals (when reports on a given topic have appeared in more than one publication). Journals, like authors, can be biased—in other words, prone to promoting a single perspective. Appearance of corroborative studies in several journals is consistent with science's requirement for reproducibility. All other things being equal, the scientific power of six studies published in as many different journals would be greater than six articles all published in the same journal. Therefore, your analysis of a study should include a quick assessment of the number of journals from which articles are cited.

Finally, look carefully at the bibliography to see if the summarized literature is up to date. A good review will refer to important articles from the past so that the reader can get a sense of historical perspective on the research issue, but it will cite the most recent studies on the same topic. Consequently, check the publication dates on references listed at the end of the article. Relatively recent publications should be mentioned unless the topic has not been researched for some time—a fact that probably merits some explanation in the literature review.

Methods (Methodology)

Good articles explain how the study was conducted; that is, they describe the method employed to conduct the study. Again, the reason is reproducibility. If an article fails to give you enough information to recreate the study on your own, it fails one of the most important tests of scientific literature. Alternatively, an article can refer you to a previously published study where the experiment was described in detail. This practice, increasingly common due to the high costs of publishing scientific journals, is acceptable if the previous study is readily accessible. Alternatively, and less acceptably, authors occasionally offer to send the information upon request. As generous as the offer may seem, I do not like it because the process takes time and is unreliable. (My students who have sent away for promised information about method have generally been disappointed with the response. References to Web sites that no longer exist are equally frustrating.)

The ability to present a sufficiently detailed description of methodology is one of the most important skills of scientific writing. You should expect it in any published report that may influence your own managerial or clinical decisions. Since the methods sections should provide enough information to allow replication of the study (or allow you to retrieve it without unreasonable effort), it will not necessarily be concise. The criteria for evaluating research methodology will be described in detail in the following chapter.

If the method is not adequately described, you are wise to doubt the study and its reported conclusions. The authors may have something to hide, such as sloppy research techniques or flimsy assumptions that would not withstand scrutiny if they were described. On the other hand, the authors may be good methodologists and bad writers. Some of the fault lies with the journal editor in this situation, but the study is still flawed.

Data

The quantitative information used to support a study's conclusions may need to be broken down and reported in several different sections of an article, so you will not necessarily find all the numbers presented under a single "Data" section. However, you should always find some data in a study. The absence of basic data effectively amounts to the author's saying, "Trust me." Don't!

The interpretation of data can vary with perspective or experience. For example, a researcher who conducts a drug experiment on mice, a clinician who prescribes the drug to humans, and a health economist who looks at the cost-effectiveness of the drug to society can all reach legitimately different conclusions about the implication of the same research finding. However, the clinician and the health economist cannot interpret the study from their different perspectives if the researcher does not present any data in the published report.

The absence of key data can be a sign of serious problems with a study, ranging from the authors failing to recognize the possibility of different interpretations of the numbers (uninformed judgment) to concealing obvious discrepancies between what the data really said and what they are reported to have said (outright deceit). Even when the analysis is done correctly and honestly, you should be cautious about accepting researchers' value judgments about data you cannot review for yourself if you are so inclined. A small difference to the researchers might be a big difference to you (or vice versa).

The numbers should be allowed to speak for themselves. At a minimum, expect an article to report the basic descriptive statistics (the subject of Chapter 5) and the appropriate test statistics (Chapters 6, 7, and 8) as you decide whether to let a study influence your own decisions. Actual, objective values of these summary measures—not subjective verbal descriptions—will be included in a scientific study worthy of your attention.

Results and Discussion

The concluding section in the standard format for scientific publication is a presentation of the results and a discussion of the findings. (Some journals present "Results" and "Discussion" as two separate sections, a perfectly acceptable practice.) This content is almost always included in a study. After all, results and discussion are probably the elements that define a study in most people's minds, so your assessment will normally focus on how well—not whether—the results are presented and discussed.

The results of a study should be reported clearly and succinctly. The section normally need not be long, and it must be understandable. A long and confusing presentation of results is often a sign that the researchers tried to address too many issues. A poor presentation of results can also indicate that the researchers strayed from the problem they originally set out to address. Therefore, make sure that the problem statement at the beginning of the article and the results at the end are consistent. You may be surprised to discover that inconsistency is not all that uncommon. You should be cautious about basing your own decisions on a study that falls into this trap because the authors quite possibly did not know what they were doing.

The discussion of results should also make sense, which does not necessarily mean that you will agree with the results. Remember that science is open-minded. Do not disregard a study simply because you disagree with the discussion of its conclusion. Indeed, you will hopefully come to appreciate really good studies that change your own thinking.

Finally, the discussion should include an honest review of the study's shortcomings: things that went wrong during the experiment, relevant omissions that were discovered after the study was completed, measurement errors that were unexpectedly introduced during the data-collection process, research subjects who failed to cooperate with the established procedures, and so on. If a study is potentially important to you, spend some time listing significant shortcomings that might have been missed—or at least, not reported—by the author. Your confidence in the study should normally be reduced in direct proportion to the number of concerns on your list that are not addressed in the discussion section.

Expect some variation in adherence to the standard format presented here. Journals are constantly trying new formats to save money on production costs and to increase the visual appeal of the printed page.

Electronic publication has introduced even more changes in the way studies ultimately reach the end user. Traditional journals certainly could be improved, so I anticipate many changes with considerable enthusiasm—especially as journals improve the integration of print and electronic media.

Still, changes in the way studies are published must not be allowed to distort the demands of science. The development of more user-friendly, visually appealing formats does not mean we can disregard content. Everything from "Introduction and Problem Statement" to "Results and Discussion" might soon begin to appear in new and different ways, but the evaluation criteria in this chapter will be just as important.

We must not let the medium become the message when it comes to science. We already know better than to judge a book by its cover or a Web site by its graphics. By the same line of reasoning, we should also know better than to judge a study by its appearance. Great presentation must not be allowed to divert our attention from sloppy science. The quality of the contents still matters. Let the reader beware.

2

Experimentation: The Foundation of Scientific Studies

The delivery of medical services is constantly challenged and frequently changed by new discoveries or new ideas. Researchers propose a new drug as an improved cure for a disease. Surgeons promote a new technique that supposedly corrects a physiological problem better than the traditional procedure. Policy analysts push new legislation that will presumably improve quality or increase access without increasing costs. Health benefit managers change the way their plans pay for care with the intent of reducing total costs. Management gurus claim their new technique will improve employee productivity.

The list of new possibilities goes on and on, raising real-world questions for the healthcare decision maker who might make changes based on new information.

- Does the new approach really make the reported difference?
- Will it produce the same results in my setting?
- Should a change be made here?

Not surprisingly, today's health administrator or clinician-executive is constantly confronted with the question of which changes to embrace and which to ignore. The most common place to turn for answers is published literature—the journal articles and monographs reporting studies that have been conducted on proposed changes. The results of recent studies are constantly put forward as reasons why certain things should or should not be done in the healthcare delivery system.

Chapter 1 identified the fundamental principles of science that ought to be evident in the literature, emphasizing a generally accepted publication format that helps guide our assessment of the scientific integrity of a study. Chapter 2 now completes the overview of good science by discussing experimentation, the actual process of scientific research to see if something probably makes a difference. As we will see in later chapters that deal directly with statistics, the word *probably* in the previous sentence is very important. A scientific experiment never proves anything with absolute certainty; it just suggests the extent to which an observed difference might have been explained by chance. The strength of a finding grows with reproducibility, that is, with a growing number of researchers getting the same results when conducting the same experiment.

Together, Chapter 1 and Chapter 2 provide the foundation for statistical analysis—the form and the function, to borrow a fitting phrase from architecture. Even though this is a statistics book, this chapter simply completes the first part of the joint condition set out at the beginning of Chapter 1: Good science and good statistics go together. If the science in a study is not valid, the statistical analysis does not matter. You need to judge an article on its scientific merits before you review its statistical analysis, so let us finish constructing the scientific foundations of a good study.

WHAT IS AN EXPERIMENT?

The concept of an experiment has been defined in many different ways, ranging from deeply philosophical to everyday practical. Rather than getting lost in an ethereal (although potentially interesting) discussion of the deeper meanings of experimentation, I believe that today's decision maker is adequately served by an understanding of the concept in its practical, modern form. Healthcare decision makers need to understand the experiment in its relevant, contemporary usage rather than ignore the concept because it has been subject to nuances in meaning over time and across disciplines.

Above all else, health administrators and clinician-executives need to be able to decide whether an adequate experiment is at the heart of a study claiming something makes a difference. This chapter provides a practical framework for judging the extent to which an experiment does what it is supposed to do. When a study makes research-based claims but fails to adhere adequately to the proper criteria of experimental method, serious questions are raised about the wisdom of using the study as the basis for managerial or clinical decisions. When a study does meet these criteria, the decision maker can then proceed with assessing the quality of the study's statistical analysis.

Way back in high school, when memorizing "handy-dandy" definitions seemed to be the foundation of learning, I was taught that an *experiment* is a *systematic approach to discovering unknown relationships in the world around us or to testing an ideas about those relationships.* I cannot remember who deserves credit for this particular definition, but it has served me well over the years. One of the twentieth century's most famous scientists, René Dubos, has provided a complementary focus on the dual nature of the concept:

> The experiment serves two purposes, often independent one from the other: it allows the observation of new facts, hitherto either unsuspected or not yet well defined; and it determines whether a working hypothesis fits the world of observable facts.[1]

In much the same spirit, Kuhn refers to experimentation as "the fact-gathering activities of normal science ... resolving some of its residual ambiguities and permitting the solution of problems."[2]

The different dimensions of experimentation would lead to discussion of corresponding difference in experimental method in a book on science. However, since this is a book on applied statistics for healthcare decision makers, we will focus our attention on the relevant (that is, the second) part of the preceding definitions: testing an idea about relationships, determining whether a working hypothesis fits the world of observable facts, or solving problems.

Please note a particularly important part of the first "handy-dandy" definition presented above. It states that an experiment is a systematic approach to inquiry. Experiments follow well-established, identifiable steps from start to finish. The steps in this system are presented here in conventional order to serve as a benchmark for your qualitative evaluation of research

---

[1] Quoted in W. I. B. Beveridge, *The Art of Scientific Investigation* (New York: Vintage Books, 1950), p. 19.

[2] T. S. Kuhn, *The Structure of Scientific Revolutions* (Chicago: University of Chicago Press, 1962), p. 27

articles. Studies that do not adhere to the basic system of experimentation are likely to be flawed in ways that might get you in trouble if you use them to guide your own decisions. Be cautious when you encounter a study that strays very far from the system, especially if the deviation from standard practice is not clearly identified and adequately defended.

The Hypothesis

The first formal step in the experimental process is specification of the *hypothesis*: a tentative statement about the expected outcome of an experiment and the mechanism by which it occurs. The hypothesis should be clearly and meaningfully presented. Further, it should be stated in an article's abstract and adequately discussed in the introduction and problem statement.

By long-standing and sensible tradition, the hypothesis should be stated in negative form, known as the null hypothesis and commonly abbreviated as H_0. Here are a few examples of properly phrased null hypotheses that might appear in articles of interest to healthcare decision makers:

- Capitation (fixed fee per member per month) does not influence clinicians' treatment decisions.
- The payment mechanism does not affect the outcomes of patients treated for infectious diseases.
- Consumption of fatty foods does not affect an individual's risk of coronary artery disease.
- The length of hospital stay does not improve patients' recovery from ophthalmologic surgery.
- The volume of operations performed by surgical teams does not affect patient outcomes.

The alternative hypothesis, abbreviated as H_a, is stated in positive form. For example, the H_a for the first sample H_0 above would be "Capitation influences clinicians' treatment decisions." Research should always test the null hypothesis (H_0). As will be shown in Chapter 6, the alternative hypothesis (H_a) is adopted only if the null hypothesis is rejected as a result of statistical analysis of experimental data.

Why has the nuance between H_0 and H_a persisted for so many years? Why not just test the alternative hypothesis? A sensible historical tradition and a contemporary problem have led to recent efforts to focus on the null hypothesis.

The historical reason for the convention of testing H_0 is the need for consistency among studies. The scientific concern with reproducibility is best met when studies of the same question are directly comparable, and comparability is less prone to error when all studies test the same version of the hypothesis. Imagine, for example, the situation where 10 studies of the same general issue all "proved" their research hypotheses. Wow, 10 studies in the same area, all coming to the same conclusion! Sounds like a strong preponderance of evidence in favor of the hypothesis—until closer inquiry shows that five studies tested H_0 and five tested H_a, which in reality shows an inconclusive 50:50 split in the studies.

The contemporary problem reinforcing the importance of testing null hypotheses is called *publication bias.* Shrinking budgets for research and growing rewards for discoveries have caused science to become very competitive over the past several decades, which in turn has caused researchers and journals to put more emphasis on "proving" something. Those who accepted a null hypothesis (H_0) began having difficulty finding journals that would print their studies. Sadly, publication bias in favor of alternative hypotheses (H_a) greatly hindered the dissemination of extremely important information about what does *not* work—negative knowledge. Some of the better journals have recently recognized and corrected the problem, but healthcare decision makers still need to be wary of the possibility that the literature may only include studies that found something made a difference. Journals have had a tendency to suppress equally important studies that failed to detect differences.

The null hypothesis should also be stated at the beginning of the experiment, before any data are collected and examined. Indeed, experimental design and data-collection techniques should be tailored to the hypothesis. The practice of stating the hypothesis before conducting the experiment and collecting the data, known as *a priori* specification, is important because the human mind is sharp enough to spot trends in the numbers. Researchers who look at data first run the real risk of specifying a hypothesis that simply confirms patterns already evident in the data, even though the patterns may be spurious or random.

Unfortunately, hypotheses are often poorly defined, misplaced, or missing altogether in healthcare studies. The situation seems to have been improving over the past few years due to the explicit efforts of the editors of some leading journals, but decision makers should still look carefully at the hypothesis. Any study that does not have an identifiable, clearly stated *a priori* hypothesis is suspect from the start. (Hopefully, it will be stated as a null hypothesis.)

A considerable amount of work should be done between specifying the hypothesis and conducting the actual experiment. This work—from conducting the literature review to planning the research method and making financial arrangements to conduct the experiment in an acceptable manner—will be largely transparent to the person reading a study's final report. Be thankful that you are spared the minute details of managing research, but be watchful for any hints that shortcuts may have been taken. If you want or need to know more about specific details that are not worthy of journal space (for example, How much did the survey cost? How was patient confidentiality protected when the medical records were reviewed by the research assistant?), get in touch with the principal investigator. Researchers tend to be friendly, albeit quiet, people who do not mind returning phone calls.

The Experiment

The hypothesis raises a question. The experiment provides *an* answer, not *the* answer (due to the requirement of reproducibility). Like symphonies, experiments come in several forms, but all are variations on a model that has been around for a few hundred years. The basic and most important form of scientific experiment—commonly represented in the world of healthcare by the randomized controlled trial, or RCT—is presented in this section as a benchmark for the evaluation of studies that might influence your decisions regarding the allocation of scarce resources in a competitive market.

Do not forget a crucial point from the beginning of this book if you are wondering why you should care about the scientific correctness of a study's experiment. The quality of research to see if something made a difference was not nearly as important to a decision maker's success back in the days of fee-for-service reimbursement when third-party payers wrote checks that covered all costs retrospectively, including the costs of bad decisions. In today's era of limited resources and accountability, administrative or clinical decisions based on flawed studies are expensive.

Using bad information can result in what is commonly described as a CLA (career-limiting act). Knowing how to spot a bad experiment might help you avoid a CLA.

The general model of an experiment has six distinct steps. If they are not followed, the validity of the study is cast in doubt because something other than the experimental effect may account for any differences found in the research. Therefore, you need to be familiar with the six steps in an experiment in order to evaluate the integrity of an experiment. I tell my students to memorize them and learn how to apply them. (In other words, if you were using this book for my course, this material would very likely be on the exam.)

1. Randomly Select the Study Sample from the Population

The *population* is the entire, defined group of people, institutions, or other entities that might be changed by the experimental effect specified in the hypothesis. For example, a population might be all the doctors in staff model health maintenance organizations in California, all the persons in the United States with hypertension, all registered nurses working in teaching hospitals, and so on. The population should be carefully defined in the written report of the study. It defines the realm in which an experimental effect may be appropriate for everyday use if the experiment produces a desired difference greater than what could be explained by chance.

A *sample* is a *representative* subset of a population. Research is conducted on samples when testing every single member of the population would cost too much money or take too much time, and statistics was developed to meet the special conditions created by the common practice of conducting research on samples. Technically speaking, statistics is not necessary when a whole population is being tested because inferences from a sample to a population are not necessary.

Random selection is the acceptable method for making sure that the sample is a representative subset of the population. Populations can have a lot of built-in diversity in terms of age, sex, race, income, physiology, health status (physical and mental), health history, education, insurance, and other factors that might interact with an experimental effect. The diversity with respect to these factors in the sample groups needs to be comparable to the corresponding variation in the population.

THE RANDOMIZED CONTROLLED TRIAL

The *randomized controlled trial* (RCT) is in some ways the scientific equivalent of a trial in our judicial system. An experiment effectively puts a hypothesis on trial. Evidence is gathered and weighed to see if the proposed relationship between cause and effect is demonstrated beyond a reasonable doubt. A scientific trial, like its courtroom counterpart, has procedural rules: randomness and control. If a study falls short on its adherence to these rules, a mistrial should be declared, and the result should not be used as precedent for decisions made by healthcare administrator, clinicians, and policymakers.

Randomness is the procedural requirement governing selection of subjects for the RCT. To be *random*, each possible outcome of the process (for example, each person in the population being studied) must have an equal chance of being selected for participation in the research trial, and each act of selection much be independent of the others. In other words, no test subject should be allowed in the experiment on any basis other than luck of the draw. Chance is the only factor that fully meets the requirement of randomness.

The *control* in an RCT is the set of factors that isolate the experimental effect, minimizing—or, preferably, eliminating—the possibility that any other factor might explain the outcome of the trial. (The *experimental effect* is the change being tested by the hypothesis, such as the new drug, the modification in a surgical procedure, the different payment system, the modified staff pattern, and so forth.) The concept of control requires that everything else be the same so that the only variation is the experimental effect. It is science's equivalent of economics' *ceteris paribus*, all other things being equal. The difference is that economists assume it; scientists require it.

Nonrandom methods are quite likely to produce a study sample that does not represent the population, which in turn produces bad research. Letting people volunteer for an experiment, a process known as self-selection, is a particularly good example of a potential problem. Individuals who believe they will benefit from an experimental effect (for example, a new drug being tested) are much more likely to sign up for the study than those who do not expect to be helped. Consequently, studies conducted on a self-selected sample will not truly test the effect on the entire population.

Random selection does not guarantee a perfectly balanced sample, but it is not subject to the potential for bias inherent in the nonrandom alternatives. Randomness leaves the selection to chance, not human judgment with its potential for bias. When used to produce a sample of sufficient size, random selection from the population is an essential foundation to good research. (The important and related topic of sample size is discussed in Chapter 4.) The more the sample selection process deviates from randomness, the less you should act on the basis of the study's results.

2. Randomly Assign the Study Sample to Experimental and Control Groups

Once the representative sample has been selected from the population, its subjects must then be divided into groups so that an experiment can be conducted. The number of groups depends on the nature of the research. If only one experimental effect is being tested, two groups need to be created. One group, called the *experimental group,* will be created to receive the experimental effect. The other group, called the *control group,* will be treated like the experimental group in every respect except one; it will not receive the experimental effect.

The study sample needs to be extended to more than two groups if the trial involves more than one experimental effect. For example, if researchers wanted to investigate whether quality of care was changed by three new nurse staffing models, four groups would have to be created: three experimental groups and the control group (the "no-change," business-as-usual nursing model). The same outcome would be required to test the impact of different doses of a drug, different levels of patient participation in payment for care, different levels of experience in performing a surgical procedure, and so on. Always look for a control group, plus a number of experimental groups equal to the number of experimental effects being tested.

Assignment of subjects from the study sample to the experimental groups must be made randomly for the very same reasons that random selection must be used to draw subjects from the population. The subjects in the experimental and control groups need to represent the population, and the experimental effect should be the only difference between the groups. Random assignment is the only approach consistent with the assumption underlying statistical analysis—that the outcome might be explained by the luck of the draw. Any assignment process that takes chance out of the picture correspondingly violates the assumption that makes statistics appropriate for analyzing the study's data, so make sure that a study's sampling from the population and assignment to research groups were done randomly.

3. Establish Pretest (Baseline) Measures for Experimental and Control Groups

Since statistical analysis was developed to evaluate the likelihood that an observed difference is explained by an experimental effect rather than by chance, decision makers want to be as certain as possible that any difference between pretest and posttest measures occurred as a result of the experiment. Randomness is our best protection against finding differences that have nothing to do with the experiment, but it is not perfect. Fortunately, we have an easy way to detect this potential problem: establishing baseline measures of all groups before the experiment.

Taking pretest measurements provides built-in protection against the possibility of confusing intergroup differences that existed before the experiment with posttest differences that likely occurred as a result of the experiment. The logical reason for taking this step is pretty obvious, but many researchers either (1) fail to mention it in their published reports or (2) do not bother taking the baseline measurements. The first situation can be a sign of sloppy writing that may force you to call the researchers if the study's findings are potentially important to you. The second situation is enough to invalidate the study because the researchers simply do not have enough information to know if posttest differences between groups already existed at the beginning of the experiment.

Establishing baselines also forces researchers to address measurement in advance of their experiments. Data collection and measurement should be defined in the early stages of research for much the same reasons that the hypothesis should be specified before the experiment. Waiting until after the experiment to decide what to measure and how to measure is a common flaw, so watch out for it. On the other hand, the existence of baseline measurements is a good sign.

4. Administer an Experimental Effect to an Experimental Group Only

This simple step is the crux of an experiment, and it should be self-explanatory. The key considerations are to make sure that the control group goes on about business as usual (that is, nothing changes for its subjects) and that the experimental effect is the only difference between the experimental group and the control group.

Of course, the experimental effect should also be applied equally to all members of the experimental group. A study's published report should include enough information to make you confident that these things were done. Judge accordingly.

5. Establish Posttest Measures for Experimental and Control Groups

In Step 5, the same pretest measurements that were taken in Step 3 need to be taken after the experimental effect is administered in Step 4. This step is also self-explanatory, so detailed discussion should not be necessary here. However, make sure that the posttest measurements were made in strict accord with the same procedures that were used to collect the pretest numbers with which they will finally be compared in Step 6.

Any changes in data-collection techniques or measurement between Step 3 and Step 5 can give a false appearance of experimentally induced differences. In this situation, such differences are explained by the data-gathering process rather than the experimental effect. Researchers will occasionally use someone else's measurements (such as U.S. Census or state health department data) for the baseline and collect their own post-test data. This practice can easily introduce a measurement error that invalidates the study.

6. Analyze Experimental Data for Statistical Significance

Statistics, at last! The use of statistical analysis is the final step in a research trial. It is not appropriate in research studies unless all the preceding steps in the experimental process have been properly executed. Chapters 5 through 8 describe in detail how statistics should be applied to specific tasks of analyzing experimental data, so the important point here is to make sure that statistics is put in its proper place—at the end of the experiment.

Having established the essential scientific foundations of quantitative analysis, we are now ready to move on to the "real" stuff of statistics: data. This is where most statistics books start, but as the first section of this book has shown, the numbers are meaningless if they are derived from inappropriate applications of scientific method or experimental design. I trust that the very different starting point of this book has proven worthwhile. You now know what questions to ask and what answers to require in your decision about whether a study is of sufficient quality to influence your activities as a healthcare decision maker.

EVALUATION STUDIES

A scientific experiment is the appropriate method for assessing the likelihood that an identifiable experimental effect explains an observed difference. It is the best tool we have for investigating specific cause-effect relationships. However, some people are interested in investigating the relationship between programmatic interventions (for example, antismoking education, stress management counseling, physician peer review) and changes in a targeted population. This form of inquiry is commonly called program evaluation or evaluation research.

Evaluative studies are designed to measure a program's relative success in meeting predetermined program objectives. Measuring the attainment of objectives is not the same thing as testing a hypothesis, and the procedures for evaluating a program are much less rigorous than the comparable steps in the scientific method. In particular, evaluative studies are not controlled. They can show the extent to which programmatic goals were met, but they cannot demonstrate that the program made a difference in the scientific sense. Program evaluation can be done very well and can be quite useful, but it should not be confused with scientific research.

Scientific research starts with a hypothesis. Evaluative studies start with the identification of measurable objectives for the program. Evaluators then define quantitative and qualitative criteria for measuring the objectives, take appropriate baseline measurements, and make the same measurements at periodic intervals (such as at the end of each program year).

The program's success is measured by the extent to which the program's participants performed as desired.

For example, a smoking cessation program aiming for a 25 percent 1-year success rate would be evaluated quite favorably if 32 percent of the participants had actually quit smoking at the end of the first 12 months. On the other hand, the evaluation of the 32 percent "success" rate would not be favorable if the program's predetermined goal had been to get everyone (100 percent) to quit smoking. In either case, the actual causes of observed changes cannot be explained with a degree of statistical confidence because evaluation is not a controlled experiment.

Section II

The Fundamental Importance of Data

Governments are very keen on amassing statistics. They collect them, add them, raise them to the *n*th power, take the cube root, and prepare wonderful diagrams. But you must never forget that every one of these figures comes in the first instance from the village watchman, who puts down whatever he damn pleases.

Sir Josiah Stamp, 1896
Inland Revenue Department (England)

3

Numbers Good and Bad: How to Judge the Quality of Data

The first section of this book began with the proposition that good science and good statistics *must* go hand in hand. Its two chapters provided a framework for evaluating the scientific integrity of procedures used to collect data for the types of studies that healthcare executives and caregivers typically consult when deciding whether a certain decision might make a desired difference in organizational or clinical performance. A common theme appeared in the discussions of both scientific method and experimental design: Good statistics cannot be used to compensate for bad science. To borrow an old metaphor that makes the same point in considerably more picturesque language, you cannot make a silk purse out of a sow's ear.

The same point holds equally true when it comes to the numbers that get crunched and made into studies. Good data and good statistics *must* go hand in hand if a study is to be of any value; good statistical analysis cannot be used to compensate for bad data. The quality of the numbers is a separate issue from the quality of the method used to collect them, so this chapter takes a careful look at the characteristics of good data and provides helpful hints for evaluating them. (By the way, *data* is a plural word; *datum* is the singular form. Get used to saying, "Data are …" It shows you are sophisticated. Well-edited publications use the word with proper subject–verb agreement.)

LETTING THE NUMBERS SPEAK FOR THEMSELVES

We Americans have a fascination with data that I have not found in other countries. Numbers have remarkable presence and power in our daily lives. From "factoids" on *CNN Headline News* to the multicolored charts in *USA Today*, from C-Span's unadorned coverage of expert testimony at congressional hearings to player and team stats crammed onto the sports pages, data are at the core of our daily experiences. (Indeed, more than one sportswriter has argued that baseball is our national pastime precisely because it produces so many statistics. Where else but the United States could "Rotisserie Leagues" attract as much attention as the game itself?) A day without numbers seems almost, well, un-American.

Unfortunately, we are not very discriminating when it comes to accepting data. We implicitly assume that numbers are correct, that numbers would not lie. We tend to accept data as reality. We hardly ever stop to ponder the fact that data are nothing more than arbitrarily defined measurements of the world around us. They cannot be any more accurate than the ruler or the scale or the sensor or the thermometer or the counting device that was used to take the measurements. And even when the measurement is precise, the data that ultimately come to a researcher's or decision maker's attention are no more accurate than the meter reader or the typist or the proofreader who was responsible for transferring the numbers from the measuring instrument to the data-reporting system. Fortunately, the automated transfer of digital data from measuring devices to computerized data bases effectively eliminates transcription errors, but it still does not guarantee that the data are appropriate measures of the phenomena being observed.

Obviously, data error can be introduced into the process at any number of points along the way. The quantitative information that ultimately crosses our desks is not necessarily accurate, so we should be concerned about the quality of data before we base decisions on them. Data users need to have their own version of the computer programmer's aphorism, "Garbage in, garbage out" (GIGO). Bad data in, bad statistics out, or something like that. (Please get in touch with me if you come up with a more elegant or catchy way to express this fundamental truth.) To operationalize the importance of good numbers, let us develop a user-friendly discussion of issues concerning the quality of data.

OF GRAPES AND GRAINS

Two "sentinel events" in my life have sparked a career-long interest in the quality of data. Note that both these experiences took place outside the United States. Residents of other countries seem to have a naturally healthy skepticism concerning data, perhaps because quantitative information has not been trivialized by overuse.

My first exposure to flawed data happened when I was taking an international economics seminar at the University of Paris. I was in a group of four students assigned the task of bringing together in one source the available data on a neighboring country that did not publish its economic statistics in any single document. Curiously, we discovered that the country's reported exports of wine exceeded reported production. One of the numbers had to be wrong because a country could not ship more wine than it could produce, *n'est-ce pas*? Well, a reporter heard about our finding and looked further into the question. He created an international scandal by discovering and reporting that both numbers were correct, but the "wine" being shipped abroad contained a lot more than grape juice (that is, water, oil, and other liquids too unpleasant to mention). Although the measurements were correct, I quickly learned that the definitions of data were just as important as the measurements themselves. Domestic wine and exported wine were not the same thing, even though they were sold in the same liter bottles.

A few years later, at the University of Geneva, I studied with a professor who was an expert in the United Nations' databases. He had become suspicious of the field crop yields being reported by some developing countries, so he fashioned an indestructible tub that held exactly one bushel of grain and spent a summer visiting agricultural reporting stations in Africa and Asia to compare his accurate vessel with the volume of the bushel measures used locally. He found such a wide discrepancy in this presumably standardized measurement unit that many of the UN's agricultural statistics were declared meaningless—even though government planners and grain traders had been relying on the data for years. I learned that a bushel is not a bushel is not a bushel.

WHAT ARE DATA?

Generally speaking, data are the stuff that makes statistics possible and necessary. Statistics was not even invented until governments started collecting lots of data in the eighteenth century. (The etymological root of *statistics* is *state,* as in political unit or nation–state. The word itself reflects statistics' origin as a tool to help states—that is, governments—manage an ever-growing volume of numbers that measured the population and the economy.)

Data are the recorded and reported measures of real-world phenomena—the values of variables, which are the possible numeric (that is, quantitative) measures of an event or experiment. As such, data can be no more accurate than the instruments and the systems used for recording and reporting them. To accept data as truth is to assume implicitly that the people who compile them are careful and honest. Because the recording and reporting of data involve humans, we ought to know better than to accept numbers without reservation. We need to think of data as more than just numbers.

Data by themselves are not usually very useful. They need to be seen as one of the links in a chain of activities that helps us make the right (or, under the circumstances, the best possible) decisions. Many philosophers of science have written about the chain of events that links thought to action in the quantitative world. Harlan Cleveland has summarized this reasoning in a five-part, interactive process[1] I use as the basis for a conceptual model that helps put data in their rightful place:

1. *Thought*: The initial stage of quantification, characterized by conjecture, curiosity, hypothesis, and so on.
2. *Data*: Thoughts quantified to become "undigested observations, unvarnished facts."
3. *Information*: Numbers put into useful form, "organized data."
4. *Knowledge*: "Organized, internalized information."
5. *Wisdom*: "Integrated knowledge … which can be used to do something," such as make good decisions.

[1] Harland Cleveland, *The Knowledge Executive: Leadership in an Information Society* (New York: E. P. Dutton, 1985), pp. 22–23.

The important point of this model is that data are only part of the big picture. Careful thinking is an essential first step that should be taken before data are collected in the second step. Neither of these steps is covered in traditional statistics texts or courses, but both are essential foundations for proper applications of statistics. Descriptive statistics (Chapter 5) is the tool for the third step in this model, organizing data to create information. Inferential statistics (Chapter 6), relational statistics (Chapter 7), or explanatory statistics (Chapter 8) are then used in the fourth step to turn information into useful knowledge.

Do not get lost in philosophical nuances of the difference between the fourth and fifth steps: knowledge and wisdom. Wisdom is the art of being a good, experienced decision maker. Statistical analysis of data can give you knowledge, but it cannot make you wise. And if you want to use data wisely, you need to be very sensitive to two *very important* (that is, these concepts will be on the exam) measures of the quality of data: validity and reliability.

Validity

Validity refers to the *meaningfulness of data, the extent to which the variable being measured truly represents the underlying thing or phenomenon that is under investigation*. I know this definition sounds a bit philosophical, but validity of data is analogous to the concept of essence that I learned back in my undergraduate philosophy class. To a philosopher, essence is the fundamental nature of something as conveyed by its name. The word *chair*, for example, conveys a very specific image in the mind's eye. A chair is something to sit on, but not everything you can sit on is a chair. You can sit on an upturned garbage can, but it does not have the essence of a chair. The essence, or validity, of "chairness" comes from having four legs and a seat and a back, all in general proportions that are relatively comfortable to a seated person.

Valid data are those pieces of information that meaningfully represent the subject of a study, just as a word can define the essence of an object. Suppose you want to collect data to learn more about teen pregnancy in order to decide whether a certain intervention is likely to make a difference. How do you measure teen pregnancy? Well, studies commonly measure the magnitude of the problem with the reported number of babies born to unmarried mothers under 18 years of age, a statistic available from the local health department. Is it valid? No way! It measures only pregnancies brought to term.

Terminated pregnancies are part of the problem but are not included in the number commonly used to measure it. The statistic further fails to capture the unwanted births to teen moms who got married to avoid social stigmatization. It also misses the local girl who gets pregnant but goes to another data unit (city or state) to have her baby. When we use the out-of-wedlock number, we only measure part of the problem—which provides an invalid picture of the whole problem.

Once sensitized to the issue of validity, you will find—as I frequently do—that many studies are themselves invalidated because they rely on data that are imprecise measures of the phenomenon under investigation. For example, the payment rates of diagnosis-related groups (DRG) have been used as indicators of the cost of providing a hospital service, which is misleading because DRGs have never been set on the basis of actual production costs. Malpractice complaints have been used to measure quality of care, which is invalid because most actual cases of negligent practice never end up in court and because most malpractice suits that go to trial are decided in favor of the doctor. Patients' hospital records are often used to provide data about the quantity of care, but hospital records can give an incomplete picture because they do not always capture related outpatient services, care purchased from hospitals or doctors outside the data-gathering system, or services obtained from alternative providers (such as massage therapists or acupuncturists) whose data never enter the system.

Consequently, as you review studies that might influence your managerial or clinical decisions, spend a few moments thinking about the validity of the data being used in a study. Ask yourself if the variable that was measured is a meaningful proxy for the problem under investigation. If the study uses data from an apple orchard to make statistically significant statements about orange groves … well, you get the point.

Reliability

Reliability refers to *the accuracy of measurement, the repeatable precision of the quantification process.* We tend to assume that researchers always make sure their data are accurate. They do not, so we should not. We need to look carefully for signs that the data are potentially inaccurate (even though they might be extremely valid).

Measurement error can be a serious problem in one of several different ways. First, researchers have been known to use uncalibrated instruments—balance scales that have not been checked for years, thermometers with separated columns of mercury, computer files that have never been verified, and so on. The problem is particularly serious in multisite or time-series trials when instruments are not calibrated at all data-gathering locations and at the beginning and end of the study. Imagine a two-year, multisite study that uses patient weight as a measure of treatment effect. The results could be meaningless if the scales at one site read high and the scales at another read low or if weights at the beginning and end of the study are recorded on different, uncalibrated scales.

Even when instruments are calibrated, a second form of fatal measurement error (error that kills a study, not a patient) can occur if the calibration is not performed with respect to a national or international standard. Pathologists deal with this potential problem by using reference standards; in other words, their laboratory instruments are calibrated against samples that are certified to be accurate at specific levels of precision. Think of the reference lab when you review a study's data. You have reason to doubt the results of a study if you do not find evidence of calibration to a common standard when calibration is necessary.

A third type of problem with reliability is errors in recording data. This problem is hard to identify because *primary data* (that is, data collected specifically for the study) are almost never displayed in published reports. Here, you can give the benefit of the doubt to the authors unless you spot an obvious indication of recording error. The more the data were transformed during the study, however, the greater the opportunity that an observation was improperly read or erroneously recorded.

Problems of validity and reliability are all the more likely to occur when a study uses *secondary data* (that is, data that were collected by somebody else for another purpose). My favorite good example of a bad example is the U.S. Census and many of its component surveys related to consumer spending. Census data are definitely not calibrated. Take income statistics, for example; if you and I are both asked to report our income on the census form, you might very precisely report the salary from your job as a healthcare decision maker but fail for whatever reason to mention sizable net earnings from the rental properties you own. (Remember, this is the census, not the IRS.) On the other hand, a self-employed person never knows how much money will be made in the current year, so his or her advance statements of income will be a wild guess. Further, since we only fill out the census forms every 10 years, a bureaucrat is responsible for attempting to adjust the data to the current year. (I will not even try to discuss all the reliability problems inherent in letting someone else adjust data that were flawed from the start.)

Secondary income data from the decennial census get used in all sorts of research—such as studies of health status and income or health services utilization and income—that could be useful to decision makers. But the results are not very useful if the data are flawed. Many comparable problems can arise with numbers from any secondary source because they were not collected specifically for the purpose of the hypothesis under study. Some researchers do not even pay sufficient attention to issues of data quality when they collect primary data.

Unfortunately, the validity and reliability problems are seldom discussed in research articles, so you need to be sensitive to them and to interpret studies with an appropriately sized grain of salt. Here are a few other areas where you will generally need to judge for yourself because researchers tend not to deal fully with the potential problems.

Location Bias

Researchers tend to collect data and conduct studies close to home. Consequently, not a whole lot of research gets done in inner cities, remote rural areas, or other areas where researchers tend not to live and work. Be wary of location-specific studies that implicitly or explicitly extrapolate their results to other areas where different circumstances might lead to different results. Do not use a study in your setting if you think it is not directly comparable to the locale where the research was conducted.

Sex Sampling Bias

Although my experience suggests that the research community tends to be sexually balanced (that is, male and female researchers in proportion that does not suggest widespread discrimination on the basis of gender), research itself has focused disproportionately on white men. Federal investigations have highlighted the relative inattention to women in clinical research, and Congress takes occasional action to direct more funding to research involving the female population. Results of studies involving men are often invalid when extrapolated to women, so be attentive to the possibility of this problem. The same problem exists with respect to race and age. Members of minority groups and children are seldom represented in proportion to their numbers in the population.

Dollar Measurement Bias

Our basic monetary unit is a key variable in many studies that will cross a decision maker's desk. The dollar might seem to be about as valid and reliable as a measure could get, but it is not. A major problem with money as a measure is variation in its value (that is, how much it will purchase) over time. We all know that a dollar today buys a whole lot less than it did 10 or 20 years ago, but economists have yet to devise a good technique for expressing the dollar in some constant value. The consumer price index is the common approach to the problem of standardizing the value of the dollar over time, but it is inherently flawed (see "The Consumer Price Index").

Adjusting the dollar for differences in purchasing power in different locations at the same time is an equally vexing problem. For example, a low-income mother in the inner city and an upper-class mother in the suburbs do not get the same care for the same amount of money. The sum of $300 will purchase one visit to the downtown hospital's emergency room or three visits to a suburban pediatrician—hardly an equal purchase for the same dollar amount. Also, some people get health services without paying for them (for example, Medicaid or charity care patients), so studies of out-of-pocket medical expenditures can understate care actually received. Finally, international comparisons of healthcare spending are approximations at best, due to the imprecision introduced by the use of fluctuating exchange rates to standardize the value of different currencies.

Counting Error

The databases available to healthcare researchers present some vexing problems related to the seemingly simple measurement activity of counting. For example, subjects with more than one name (for example,

THE CONSUMER PRICE INDEX

Economists have long recognized the need to adjust time-series data for changes in the buying power of the monetary unit. The consumer price index is the best solution devised so far, but it has limitations that need to be understood by those who use it.

The foundation of a price index is a market basket of goods that is priced periodically, such as monthly or annually. The composition of the market basket is carefully defined to represent consumption patterns in the base year, and all the items in the basket are then priced. The total cost of the basket is assigned an index value of 100. In each subsequent market period, the identical market basket of goods and services is priced at current value. The total cost of the basket in each subsequent period is then expressed in terms of percent change from the base year index value of 100.

To illustrate the principle, imagine a market basket of goods that cost $60 in the base year; $60 becomes the index value of 100. At the end of the next year, the cost of the same goods has risen to $66, a 10 percent increase from the previous year. The index value of the $66 market basket at the end of this year is 110, a 10 percent increase over the base index of 100. In other words, a price index is a relative (percent) expression of differences in absolute (dollar) values over time.

The medical care component of the Consumer Price Index, commonly designated the MCPI, illustrates the major problem with price indexes. The base year for the MCPI was 1967, so the original market basket included medical goods and services that were common at the time: Tetracycline was the representative prescription drug, a hernia repair was the surgical procedure, and a house call was one of the included doctor services. Obviously, the typical medical market basket several decades later would not include these items due to dramatic changes in medical science, so the market basket needs to be changed. But changing the market basket invalidates comparisons with the base year.

Price indexes are only valid over time when the market basket is unchanged over the same period of time. Frequent and major changes in the medical market basket effectively make price indexes meaningless for long-term studies in healthcare.

maiden name and married name, name correctly and incorrectly spelled) or identification number (for example, multiple Social Security numbers) will be identified as different people in a database. Studies that rely on counts of patients would correspondingly overestimate the actual size of the population. Another common problem is created by the fact that some clinicians or patients prefer to schedule multiple visits for a multi-step therapy, while others might prefer a single visit. A study that ignored this fact might conclude that more care was provided in the former situation because the patient had more visits, but the two situations may be identical in clinical terms. The costs of standardizing the count can be substantial, so many researchers conduct studies without addressing it.

Definitional Error

In health studies, a rose is not necessarily a rose is not necessarily a rose. For example, a study comparing health services utilization under different practice arrangements could produce misleading results in the absence of procedures to standardize the definition of key variables such as a patient visit. A member of a health maintenance organization (HMO) might be seen by a nurse practitioner for the same care provided by a primary care physician under a fee-for-service plan. If only doctor contacts were counted as patient visits, the care given to the HMO member would be understated. Questions have even been raised about the validity of equating all doctor visits. Is the care provided by a subspecialist the same as the care provided by a generalist? The issue is occasionally researched, hopefully with valid and reliable data.

Adjustment Error

Some studies and many media stories about those studies fail to make relative adjustments for absolute differences in key variables when they report results. Five deaths in each of two different study sites might give the impression that there was no difference between the sites, but the difference would be highly remarkable if one were a rural community with 2,000 inhabitants and the other a city with 200,000. On the other hand, the difference would be interpreted quite differently if the five deaths occurred over two years in the small town and over two days in the city. Standardization for population, time, and other measurement standards is always important when samples are not identical, but studies do not always make the necessary adjustments. Do not be fooled by unadjusted results.

Time Measurement Error

Time would seem to be about as standardized as a measure can get, but several distortions can be found in healthcare studies. For example, be on the lookout for studies that fail to distinguish between calendar year and fiscal year. Hospital cost studies occasionally lump data into the same year for comparison purposes, even though some of the hospitals start their fiscal years as much as three quarters later than the others—close to a year's difference. A few studies of the early impact of prospective payment committed a similar error by comparing hospitals at specific points in time even though hospitals were brought into the diagnosis-related group (DRG) system in staggered fashion over several years (that is, a hospital that had been receiving DRG payments for some time could have been compared with one that was still receiving retrospective cost-based reimbursement).

The quality of data has received a lot of attention in this chapter because it is one of the most important considerations in judging the overall worth of a study. You should become familiar with these issues so that you are comfortable applying them in your own reviews of published literature. You should also be on the lookout for other data-related problems that have not been included in this chapter. With additional experience and careful thinking, you may well identify even more items that need to be added to this important list.[2]

TYPES AND LEVELS OF MEASUREMENT

Just as the world is divided into two types of people—those who divide the world into two types of people and those who do not—data are divided into

[2] Please contact me if you do develop some useful additions. I will gratefully include them in subsequent editions of this book so that others can benefit from your careful thought and observation.

A TIMELY PROBLEM WITH PRIMARY DATA

Using secondary data greatly increases the potential for problems with reliability and validity of data because data collection is not tied directly to experimental design, but using primary data—data collected specifically for a particular study—does not automatically make the problem go away.

I was once asked to read a master's of science in health administration (M.S.H.A.) thesis written by a nurse who managed the emergency department of a large, urban hospital. She had studied whether a new model of patient triage made any difference in patient outcomes in her department, a place where time is an important element in quality of care. The research design was exceptional. It had a good hypothesis, clear problem statement, nice review of the literature, well-controlled sample selection, careful research design, and the like. Her data analysis was also thorough and thoughtful. To collect the data for her study, she had prepared a data form to be filled out by caregivers.

I was impressed with everything except one aspect of her approach to data collection. Time was a key variable in her analysis, and all employees who treated a patient wrote down the beginning and ending times of their contact with the patient: for example: "started contact at 7:38 a.m. and ended contact at 7:51 a.m., for 13 minutes of care." In particular, the total duration of treatment was defined by the difference between the starting time on the first sheet and the ending time on the last. Yet, I could not find any evidence that all the caregivers had been asked to calibrate their watches during the study.

I called the student and issued a challenge: set a very accurate timepiece to Greenwich Mean Time (GMT) and compare its time with the time on every clock in the ER (such as employees' wristwatches, wall clocks, radio clocks, time indications on computer screen, and so forth) that might have been used to record the times on the data-collection sheets. I then asked her to plot the deviations from GMT.

"Ouch," she exclaimed when she called me back. Differences in the various timepieces accounted for nearly one-third of the total variation in treatment time between the experimental and control groups. She repeated the experiment with coordinated clocks, got an A+ on the revised thesis, and went on to graduate *cum laude*. To the best of my knowledge, she never did publish the study, but if it had been published in its original version, ER directors all around the country might have made a change that was based on unreliable data, even though the rest of the study would have been highly rated according to the criteria presented in this book.

two types: parametric and nonparametric. The distinction between parametric and nonparametric data is one of the most important theoretical concepts in statistics, yet it is commonly ignored or barely mentioned in statistics textbooks.[3] In instances where the distinction is barely mentioned, some authors go so far as to state that the distinction is unimportant. I disagree, and I hope you will too, after reading this section.

I am among the purists who believe that the difference between parametric and nonparametric data is important and needs to be understood by anyone who wants to use statistics correctly. Indeed, why would statisticians have spent so much time developing distinctly different theoretical approaches to the analysis of parametric and nonparametric data if there were no consequential differences between them? Parametric and nonparametric statistical tests exist because parametric and nonparametric data are different, and good researchers are careful to match the type of statistical test to the type of data. Looking for a proper match between type of test and type of data should be an important part of your assessment of a published study.

The key issue in the difference between parametric and nonparametric data is consistency in the unit of measurement. Parametric measurement is consistent from user to user; nonparametric measurement is not. Do not read this statement as a pejorative comment on nonparametric data. Both forms of measurement are extremely useful when properly analyzed. (I will admit one bias. Although I have nothing against nonparametric data

[3] Four of the nine statistics texts in my personal library do not even mention the distinction, much less discuss it. In my opinion, the distinction gets adequate discussion in only two of the five texts that do mention the differences between parametric and nonparametric measurement and its importance to statistical analysis.

and use them in my own research, I believe that parametric measurement should be used whenever possible because parametric statistical tests are more powerful than their nonparametric counterparts. Nonparametric data are very helpful when parametric data are not possible.)

Parametric Data

Parametric measurement has consistent, and therefore universal, meaning. The scales of parametric measurement have the same gradations everywhere they are used. By extension, parametric data are those observations on variables that do not have to be defined beyond being identified by name because everyone has the same understanding of the basic measurement unit and its subdivisions. Further, the distance between points on the scale is everywhere the same.

The meter is a good example of a parametric measure of distance. The meter is universally defined as one ten-millionth of the distance from the equator to the pole, measured along a meridian. (The meter has been defined even more precisely in terms of the wavelength of radiated light under specific conditions.) A platinum bar exactly one meter in length is kept at the International Bureau of Weights and Measures in Paris, and it serves as the international standard for calibrating all other meter sticks. Assuming the use of properly calibrated instruments, a research subject measuring 1.74 meters in height in Germany is exactly as tall as a subject measuring 1.74 meters in Pakistan. Authors who might be doing a study of Germans and Pakistanis do not have to define height in any way other than a meter because a meter is a meter is a meter. That is parametric.

There are similarly consistent units of measure for time, weight, volume, color, and other attributes that have been universally defined. In other words, data measured in units that speak for themselves are parametric. Two levels of parametric data, ratio scales and interval scales, are further defined. For both, the intervals between points on the scale are always the same. The difference is that a ratio scale has a true zero (such as weight, time, temperature on the Kelvin scale), whereas an interval scale does not have a true zero (such as temperature on the Fahrenheit scale). Do not worry about the difference between ratio and interval scales. Just be sure that you know how to recognize parametric data.

From the perspective of statistical theory, the neat thing about parametric data is a built-in, group-to-group consistency in the unit of measurement.

As will be shown later in the discussion of inferential statistics, a consistent measurement scale is the condition that makes theoretically possible the most powerful comparisons of distributions from experimental and control groups to see if actual differences between them are greater than any difference that might be explained by chance. (Do not worry if this point does not quite make sense now; you should be able to understand it after reading Chapters 5 and 6.) The mathematical operations of parametric statistical tests—in other words, the computations expressed in all those equations involving sums of squared deviations from the means, square roots of the sample sizes, and the like—are made possible by the fact that distributions of different samples can all be reduced to measurement along a scale that has universal meaning and consistent distances between points.

Nonparametric Data

Nonparametric measurement is defined by the user, and the definition is arbitrary. Therefore, nonparametric data do not have universal meaning, nor do they have any uniform distance between points on the measurement scale. Consequently, comparisons between groups cannot be made on a single scale because magnitudes are not necessarily consistent from group to group or even within groups. For this reason, nonparametric data are also called "distribution-free" data. (Statistics borrowed the term *parametric* from mathematics, where it refers to the specific distribution of values in a function. Therefore, *nonparametric* effectively means *not distributed*.)

Subjective assessment of consumer satisfaction is a good example of user-defined measurement. Using a common data-gathering tool called a Likert scale, customers at the ABC Health Center might be asked to respond to the following item on a questionnaire:

> How do you rate the quality of your care at ABC Health Center?
>
> _____ Bad
> _____ Poor
> _____ Fair
> _____ Good
> _____ Excellent

A consecutive number is assigned to each response (Bad = 1, Poor = 2, Fair = 3, Good = 4, and Excellent = 5) so that each patient's response can be entered into a database for computerized analysis.

Not to be outdone, the XYZ Health Center also decides to conduct a consumer satisfaction survey. It, too, uses a Likert scale on its questionnaire:

How do you rate the quality of your care at XYZ Health Center?

_____ Excellent
_____ Good
_____ Fair
_____ Poor

A consecutive number is assigned to each response (Excellent = 1, Good = 2, Fair = 3, and Poor = 4) in order to allow data analysis by computer.

Both approaches to data gathering are perfectly defensible, but you can immediately see how the user-defined nature of this process prevents meaningful comparison of the results. The ABC survey used a five-point scale going from worst to best, while the XYZ survey used a four-point scale going from best to worst. Is consumer satisfaction identical at ABC and XYZ if the average score of 100 patient responses per clinic was 4.0? Obviously not; it is good at ABC and poor at XYZ. To make any sense of the information, you need to be told the exact scale and numerical scoring method because there is no universally accepted measure of subjective information.

To further complicate matters, two patients who received absolutely identical treatment at ABC Health Center would probably give different ratings if they had different expectations. A new patient whose point of reference was the Mayo Clinic might tend to rate ABC somewhat more critically than a new patient who had received all previous care from an inner-city Medicaid mill. In the absences of a calibrated satisfaction meter that we can attach to a patient's arm like a blood pressure cuff, we must realize that a nonparametric scale does not provide a consistent measure from subject to subject.

This example clearly shows that nonparametric data are much less precise than their parametric counterparts. However, in spite of the limitations of nonparametric measurement, nonparametric data are useful, and they can be analyzed by statistical tests that use a different (vis-à-vis parametric) theoretical approach to comparison. The computational foundation of nonparametric statistical tests will be described in later chapters.

For now, the important point to remember is that nonparametric data are fundamentally different from parametric data, and they should be analyzed with the appropriate nonparametric test. Dozens of nonparametric tests have been developed to reflect the special characteristics of distribution-free data based on arbitrary, user-defined measurement.

This chapter has presented some very important information about the quality of data and the fundamental differences between data that have universal meaning and known distributions and data that do not. Studies should be judged according to these qualitative considerations. However, before exploring the statistical models that transform data into information and information into knowledge, let us look at one more important part of the big picture: data-collection processes.

4

Samples and Surveys: How Numbers Should Be Collected

Each of the three previous chapters explored a basic building block of good research: scientific method, experimentation, and data. If a study conforms acceptably well with the qualitative criteria discussed under each of these headings, its quantitative information may be suitable for statistical analysis, and—assuming good statistical practice—its findings may be worthy of a decision maker's careful consideration.

As strange as it may seem, valid and reliable data can be collected in ways that cast serious doubt on the final results of a study. Imagine, for example, researchers selecting a biased sample (for example, asking only cigarette smokers to participate in a study of attitudes toward smoking bans in public places, or choosing only hospitalized patients for a study of a community's general level of health) and then measuring the subjects' opinions or health status with the utmost care and precision.

Just as good statistics cannot overcome bad science, good measurement cannot overcome bad data collection. How the data are collected is as important as how good the data are, so let us add this fourth general area of concern to the list of criteria that need to be met before the statistical analysis is reviewed.

SAMPLE SELECTION

One of the important characteristics of good data collection, *random sampling*, has already been addressed in the context of experimental research (see Chapter 2). A sample data set is biased if some members of the population it represents were more likely than others to be included in the sample. The more a study's sampling deviates from a random process where every member of the population has an equal probability of being selected, the less meaningful the study's results are—even if awesome effort was dedicated to validity and reliability.

Acceptable approaches to random sampling are the subject of numerous books and journal articles that can be easily identified by any reader who really needs to know how to select a proper research sample. However, for the purposes of this book, being sensitive to the issue is sufficient. If a study's authors do not convince you that the sampling process was truly random, especially when you can plausibly imagine how some people in the population might have been systematically excluded from the sample, you have reason to doubt the data unless the researchers provide a convincing case for using a nonrandom procedure to select the study's sample.

Randomness requires a little help when a study needs to include subjects who are sparsely represented in the population. The desirable approach in such circumstances is called *stratified sampling*. If, for example, I wanted to study some aspect of the relationship between health status and ethnic origin in a typical rural area, a random sample of sufficient size would easily produce enough Anglos and Latinos because both groups are large and together constitute perhaps 90 percent of the population in many rural counties. However, the Asian population in an area of interest might be less than 5 percent of the total. If it is growing fast, it should be included in any study that might be used to change the local allocation of health resources. A random sample of the entire population would likely produce a disproportionately small number of Asians in the sample, so we would need to find some way to compile a list of all ethnic Asians and then *randomly* select subjects from that list to ensure including enough Asians in the study.

SAMPLE SIZE

Selecting samples the right way is an essential attribute of a good experimental study conducted to see if something makes a difference, but randomness is not the only issue that matters. The sample must also include

STATISTICAL ABBREVIATIONS

Sample size is commonly abbreviated with the lowercase letter *n*, followed by the equal sign and a number. For example, $n = 72$ means the sample included 72 subjects. An uppercase *N* refers to the size of the population from which the sample was drawn.

The conventional practice in statistics is to use an uppercase letter as the abbreviation for a population parameter and the same letter in lowercase to indicate the corresponding measure for a sample. Sometimes Greek letters are used instead of familiar letters from our Roman alphabet, but the practice of using capital letters for populations and small letters for samples is the norm in either instance.

enough subjects to be representative of the population. Consequently, the size of the sample must meet certain standards. Samples that are too small cast serious doubt on a study's findings, so some general guidelines are needed for determining the minimally acceptable sizes of research samples.

Not surprisingly, the issue of minimum sample size is directly related to the concept of randomness. Random selection is used to prevent human intervention from intentionally or unintentionally creating differences in the control and experimental groups used to test a hypothesis, but randomness has the potential to introduce a bias of its own in small samples. In other words, even a random sample that is too small can fail to produce representative data. Here is why …

The prototypical random event is the toss of a fair coin, one where either outcome (that is, a head or a tail) is equally probable and each toss is independent of every other. We do not have to be rocket scientists to expect that the number of heads and the number of tails should be equal or very close to equal over the long haul with a fair coin, but we also know from experience that the two possible events do not alternate in lockstep order. If I get a head on the first flip, I am just as likely to get a head on the second toss and on the third toss, and so on. (Imagine how useless coin flips would be if random events occurred in predictable order—head, tail, head, tail, head, tail, and so on. We might as well let the home team always start out with the football.)

Just how many times must we flip a coin before the outcome of this perfectly random event produces the expected outcome, an equal number of heads and tails? For example, is 10 flips ($n = 10$) enough? Well, if we repeatedly flip a fair coin 10 times and record the results of each trial of 10 tosses, we will discover that five heads and five tails occurs slightly less than 25 percent of the time. (Go ahead and do this exercise if you need to be convinced. Professors of large classes in introductory statistics often have each student record the results of 10 coin tosses. When the results for the whole class are analyzed, only about one student in four reports getting an equal number of heads and tails.)

If 10 tosses were enough to make the sample an accurate reflection of the population, we should find five heads and five tails occurring almost all the time, not 25 percent of the time. Consequently, $n = 10$ is not enough to approximate the predictable outcome of a random event. Flipping a coin 10 times does not consistently yield the "right" answer: an equal number of heads and tails. What about 20 tosses, 30, 40, or 50?

In other words, just how big does a random sample have to be in order to serve as an acceptable proxy for the population it represents? The conventional answer is *minimum sample sizes of 30*, all other things being equal. Therefore, when you review studies, you generally want to see at least 30 subjects in the control group and in each experimental group.

Note carefully that the expectation is 30 subjects per group, not a total of 30 subjects involved in the experiment. Since every good experiment has at least two groups—a control group and one or more experimental groups—the minimum number of subjects that needs to be included in a research experiment under normal circumstances is effectively 60. If one or more of the samples has fewer than 30 subjects, the study's results might be explained not by the experimental effect, but by the small sample size.

Why is $n = 30$ generally accepted as the minimum size of sample that overcomes the problem of small numbers? The answer lies in probability theory, and explaining it would go beyond the scope of this book. In a nutshell, a mathematical formula is used to compute the distribution of probabilities for various sample sizes. The theoretical distribution of random events converges toward an acceptable level of certainty (95 percent) when the sample size reaches 30 cases. (Personally, I am comfortable with a minimum of 30, but I much prefer samples sizes in the range of 60 to 100 because I like the extra margin of statistical power provided by larger samples.)

Although $n = 30$ is a good rule-of-thumb minimum for evaluating research studies, much larger samples are necessary under some circumstances. To make the point with a common metaphor, a needle in a haystack is quite unlikely to be found among the first 30 straws drawn at random. Therefore, if a study involves something very uncommon in the population, such as patients with a rare disease or persons over 100 years of age, larger samples may be needed just to ensure that the sampling procedure yields enough subjects for comparisons.

Stratification is generally a suitable solution to this problem when the study group can be readily identified for targeted sampling, but very large samples are sometimes the only defensible way to conduct a study. Be on the lookout for this possibility. Other than suggesting you think carefully about the possibility that a larger-than-normal sample is needed to ensure sufficient inclusion of something relatively rare in the population, I do not know any simple approach to determining just how large the sample needs to be. If you suspect a problem with sample size in a study that is very important to you, consult a statistician who can assess the probabilities for you.

All other things being equal, can a sample be too large? *Theoretically* speaking, no, because increasing the size of the sample is the key to increasing the *power of the test*. A statistical test becomes stronger as the sample size gets larger, that is, as the size of the sample approaches the size of the population. Indeed, statistical tests are not even necessary when the sample is the population because probabilistic inference—the principal function of statistics—is irrelevant when everything can be directly measured and compared. So, in theory, a sample cannot be too large. *Practically speaking*, however, samples can be larger than they need to be. Beyond a certain point, adding to the sample size reaches a point of diminishing marginal returns. Selecting the extra subjects becomes relatively expensive for the small increase in statistical power.

Researchers need to make trade-offs in a world where funding is tight, and sample size is just one of many factors that needs to be balanced. Some try to impress us with a large *n*, but a big sample is misleading if it is obtained at the expense of the other attributes of good research. Conversely, some researchers waste our time with studies based on samples that are too small, even when the study is practically perfect in every other respect. Samples containing at least 30 randomly selected and randomly assigned subjects are usually large enough to support the probabilistic foundation of statistical analysis. Adding to the sample size is desirable when resources allow, but enough is enough. Time, money, and other resources put into increasing the size of the sample could often be better expended on activities to improve randomness, control, precision of measurement, and the like.

RESPONSE RATE

A randomly selected sample of 30 or more subjects does not automatically confer the credibility of probability in all circumstances. It is large enough to be theoretically defensible when every randomly selected subject is entered into the study's database, but problems start to occur when selected or intended subjects have the opportunity not to be included in the database—in other words, not to respond to the invitation to join in the study. Therefore, attention to the rate of response should be one of the criteria you use when deciding whether to be swayed by reports of someone else's research.

The major problem inherent in nonresponse is the possibility that subjects who choose not to participate in a study are different from those who do respond in some way that would change the interpretation of results. If its nonparticipating subjects are no different from its participants, a study is not adversely affected by nonresponse. However, many researchers fail to realize (or realize and fail to reveal) how nonresponse may cause some very important information to be kept out of the database.

An excellent example of the potential problem can commonly be found in studies of consumer satisfaction. Imagine 120 randomly selected persons responding to a written survey sent to 200 active patients of a medical group. The sample seems good on first examination, randomly selected and well in excess of 30 subjects. Even the 60 percent response rate seems impressive ... until you start to think of reasons why 80 patients did not return the survey form. Does the missing 40 percent include all the members of minority groups, patients who perceive the clinic is so insensitive to their concerns that completing the survey would be a waste of time? Does the group of nonrespondents include all the really dissatisfied patients, people who are afraid to express their true opinions for fear of retribution on the next visit? Does it include all the visually impaired elderly who had trouble reading the form?

The list of possible reasons for nonresponse goes on and on, and that is the point. Unless researchers can describe the missing participants and present plausible arguments why nonrespondents are not different from the respondents in any way that would influence interpretation of the data, a less-than-perfect response rate is a real reason for concern. As a decision maker wondering whether to be influenced by a study, you should stop to think about any unexplained or unconvincingly explained nonresponses. Above all, be on the lookout for studies that downplay the significance of nonrespondents on the unproved assumption that they are no different from the respondents. Response rate is a significant issue. The more it falls below 100 percent, the more it needs to be addressed in the interpretation of findings.

Also, be on the lookout for perfectly balanced samples within 100 percent response rates. In my experience as a health marketing consultant, I have on several occasions been given the results of "random" community surveys to assess unmet needs, consumer satisfaction, and competitors. I am very suspicious when the sample includes exact multiples of 100 persons—and even more dubious when the sample is equally divided between men and women, users and nonusers, insured and uninsured, and so on. This seemingly perfect outcome results from a deceptive practice used in the survey research business. The survey participants are actually drawn at random, but the surveyors keep sampling until they achieve the desired mix of respondent characteristics. What they do not tell you, to create an illustrative example, is that 900 people were contacted to produce the "perfect" sample of 500, a 55 percent response rate. So what about the 400 persons (45 percent) who refused to participate in the study or were deleted from the sample because their category was already filled?

The research community does not have a commonly accepted threshold for deciding when a response rate is a significant problem. My personal experience makes me very uncomfortable with unexplained response rates below 80 percent, and even that number varies up or down with the circumstances (that is, with other factors that I think might be relevant). However, I do have serious reservations about the large number of published studies that draw conclusions with response rates below 60 percent.

Rather than provide a minimum response rate that might give you a false sense of security, I urge you to think carefully about the possible reasons for nonresponse and to interpret results in consideration of related reasons that might lead to different interpretations of a study's data. After grappling with this issue a few times, you may be ready to join me in wishing that researchers would put more money into getting high response rates from small ($n \geq 30$) samples rather than low response rates from large samples.

INFORMATION SYSTEMS

The information system itself can become a factor in defining the quality of data in a study. Ideally, the system used to collect, store, and report information should be neutral, keeping the data as pure as possible. However, data are handled with varying degrees of care, and you need to be aware of the fact that some apparent findings are the result of problems with the information system.

I first became sensitized to this problem in the mid-1980s when a state health department hired me to evaluate the method used to gather information about infectious diseases. The information was collected on a monthly basis from physicians' offices by the county health departments; the state office then compiled the reports from the counties. I visited the local health department offices in six counties, ranging from remote rural to major urban. Well, to make a long story short, I found six totally different approaches to data collection. At one extreme, the county sanitarian placed telephone calls to friends at a few doctors' offices, multiplied their estimates of infectious disease visits by a factor based on "experience," and filled in the forms that were then sent to the state. At the other extreme, the county nurse visited randomly selected offices, reviewed patient records, and estimated the overall rate using a formula from the Centers for Disease Control (CDC) in Atlanta before sending the data to the state system.

Everyone was trying to do the best job with the resources at hand, but the county-by-county approaches to data collection were so different that the centrally reported numbers were clearly not comparable. My analysis caught the attention of the health department in another state, which invited me to conduct the same analysis there. The results were identical. The information in the state database was collected and reported in several different ways, raising serious questions about the consistency of the data. The lesson: central databases can be misleading if they do not include clearly understood and uniformly applied mechanisms to ensure the comparability of data from different sources. When studies use data from a central source that has not implemented a standard system, differences attributed to an experimental effect might be nothing more than differences in data reporting. Please note that the problem is comparability, not necessarily validity and reliability. The different suppliers of information could all be making meaningful and accurate measurements, but they are not necessarily measuring the same thing.

SURVEY RESEARCH

Although this book has included some comments about survey research, most of the discussion so far has pertained to time-series data made publicly available by large organizations such as the Department of Health and Human Services, the Center for Medicare and Medicaid Services (CMS), the National Institutes of Health, state health departments, the Census Bureau, hospital associations, medical societies, and insurance industry groups. A considerable amount of the research of interest to healthcare decision makers uses the databases that these national organizations have been publishing periodically for the past several decades.

An interesting historical fact is worth noting here. Very little information was collected by any of these groups prior to the 1960s because healthcare in the United States was an issue between doctor and patient; in other words, it was "none of the government's business." Epidemiological data about specific diseases were just about the only measurements taken on an annual basis until healthcare became the government's business with the creation of Medicare and Medicaid in 1965. Consequently, most nonclinical figures for years prior to the late 1960s—data on medical expenditures, health services utilization, professional personnel, and the like—are retrospective estimates, not information actually collected at the time. They are subject to the problem of reconstructing data many years after the fact, so be cautious when reviewing studies that use older data.

Also, be attentive to potential problems introduced when reporting agencies change the way they define and measure key variables. For example, as part of an extensive analysis of medical practice costs that I conducted in 2008, I reviewed several decades of data collected by government agencies and professional organizations. The definitions and measurement of practice costs were changed several times during this period, which raised serious questions about common uses of the data in research articles and policy papers. Unfortunately, the changes were not obvious in the tables where the data were presented. I had to read footnotes and appendixes to identify issues that raised serious questions about the common practice of comparing the practice cost data over time.

Gathering data from survey research—conducting interviews with or administering questionnaires to the research subjects themselves—has become a common alternative to using numbers from institutional databases. In the not-too-distant past, studies tended to rely mostly on data *about* subjects: patients, providers, purchasers, and other participants in the healthcare system. Survey research has great potential for providing decision makers with different and better information about the behavioral dimensions of healthcare delivery. Asking consumers or providers to explain their attitudes and behaviors is at least as sensible as trying to construct explanations from descriptive data such as income, sex, ethnic status, and education.

A carefully conducted survey is a powerful research tool, but like the data-collection methods described in the previous chapter, surveys are subject to problems that can weaken the findings of studies based on their data. This chapter concludes with a discussion of key methodological issues pertaining to survey research. The list is by no means exhaustive, but it does address the problems I have encountered most often in reviewing the literature aimed at healthcare managers and clinician-executives. Consider these issues when you are deciding how much credibility to give to a survey-based study. And do not hesitate to seek additional information if your own common sense suggests problems above and beyond those mentioned here.

Survey Bias

Compared to the impersonal process of collecting information from individual or organizational records, the personalized approach of surveys can itself be a source of bias, that is, a systematic distortion of the results. For example, a person being surveyed might be less than truthful in answering questions if he or she feels that specific responses might be made available to a health professional or personal friend who works for the organizational sponsor of the survey. Fear of retribution or other adverse consequences is known to distort survey responses, leading in particular to the suppression of negative perspectives.

The existence of credible mechanisms to ensure each respondent's *anonymity* and/or *confidentiality* is extremely important in surveys where respondents might believe a consequence could result from the information they provide to a surveyor. (If a survey is anonymous, the information will be reported but the survey's sponsors will not have any way to trace it to individual respondents. If a survey is confidential, the identities of the participants are revealed, but the information they provided cannot be linked to them.) To give appropriate protections, the use of outside surveyors is often necessary in interviews where respondents might be asked to provide controversial or embarrassing information.

Therefore, your own evaluations of studies involving survey data should look for bias that might be introduced by the relatively personal nature of the process. If you encounter a survey that would make you think twice about giving honest answers, you can safely assume that unreliable information was provided by many of the people who answered the questions. That you would not want to base your own decisions on such a study goes without saying.

Respondent Objectivity

Sadly, the information provided by people who respond to surveys is not always accurate. Flaws in the data system are not the only source of inaccuracy. People have been known to misunderstand survey questions, to embellish responses, even to lie more often than researchers would like to admit. Therefore, your assessment of a study needs to address the possibility that survey data may be flawed to the point that the value of the study itself is diminished.

Social stigmatization can have a major influence on the objective value of survey data. People responding to surveys do not like to be embarrassed by giving answers that conflict with social expectations, so the wording of survey questions is very important. For example, people know that smoking is no longer acceptable in many social circles (especially those inhabited by health professionals who conduct surveys!), so many smokers will respond in the negative when asked, "Do you smoke?" On the other hand, they will be disarmed by and much less likely to lie when the question is asked, "How many cigarettes did you smoke yesterday?" Likewise, dental researchers have known for years that three times (that is, once after every meal) is by far the most common answer to the question, "How many times do you brush your teeth each day?" The estimated frequency of toothbrushing drops considerably when people are asked, "When was the last time you brushed your teeth?"

Similar problems occur with distressing (or amusing, depending on your perspective) regularity in political surveys. For example, after-the-fact examination of surveys about healthcare reform shows why so many politicians overestimate public support for proposed changes. Questions were generally worded to elicit expressions of discontent with the current system in comparison with various ideal systems already proposed by politicians. Few polls asked the relevant questions: "Do you trust the government to make the proposed change?" and "How much more are you willing to pay for the proposed change in healthcare?"

Wording the questions is a skill that separates good survey researchers from bad ones. By extension, differences in wording can make the difference between good and bad surveys because respondents' objectivity is directly affected by how questions are phrased. Competent survey researchers also know how to use several different approaches to the same general questions for *internal validation* of responses. When skillfully managed, the inclusion of differently worded questions in different parts of a survey can directly assess the overall objectivity of the responses.[1] Unless you have formally studied survey research, you probably do not have the skills to develop a good survey, but you can identify a lot of bad ones simply by thinking carefully about the questions that were asked in a study.

One final note on this very important point: Good surveys are pretested, usually several times, before they are used to collect the data that are analyzed for statistical significance. All other things being equal, be favorably inclined toward a research article when the "Methodology" section describes how the survey instrument was developed, pretested, revised, and retested before being used in the final study. I know from embarrassing personal experience that the first version of a survey is usually

[1] For a classic example of the level of sophistication that can be applied to survey design, see John B. Lansing and James N. Morgan, *Economic Survey Methods* (Ann Arbor: Institute for Social Research, University of Michigan, 1971).

full of ambiguities that will distort responses. (I believe that most other researchers would agree with me, at least under the influence of truth serum.) Developing a good survey instrument can require as much effort as the rest of the study. If you have doubts about the objectivity of the data-collection instrument, you are wise to have doubts about the survey because, as the saying goes, "GIGO" (garbage in, garbage out).

Survey Format

Wording is not everything. The layout of the form or the pattern of the interview is also a very important factor that can enhance or diminish the value of data obtained from a survey. The survey instrument is seldom presented in any detail in the published report of a study, but the format is sufficiently important that you will benefit from being aware of key issues. Here are a few criteria you can apply to the extent possible. (If a survey is very important to you, request a copy of the data-collection form or the interview template. You can then address all the issues raised in this section.)

A good survey begins with an introductory explanation of the study's purpose and clear instructions for responding. The *introduction* should give the reason for the survey and identify the sponsor, and both should be honest. Many surveyors use healthcare surveys somewhat covertly, trying to conceal the real reason for the survey or the organization that will receive and use the information. In my experience, this approach is counterproductive because respondents will not give full and honest answers if they suspect the surveyor is being less than totally honest with them. This problem becomes particularly serious if the respondents have any concerns about their anonymity.

After some unpleasant graduate school experiences with using less-than-honest approaches to surveying, I have been completely open in describing the purpose and identifying the client in the hundreds of surveys I have conducted. (Indeed, I refuse to conduct surveys for potential clients who want to hide something from the respondents.) Many participants in my surveys have told me how much they appreciate the "no-secrets" approach, and they have frequently indicated they would not have shared their true opinions if they had suspected the survey had a hidden agenda. Consequently, I am extremely skeptical of the results of surveys that fail to reveal or intentionally conceal the purpose and the end user of the information.

Clear *instructions* are important, especially in written surveys. The respondents need to be told how many responses to make on each item (for example, "Check only one response" vs. "Check all applicable responses"). As simple as this issue may seem, a remarkably large number of survey respondents will provide useless information if the acceptable number of responses is not perfectly clear. An example of a properly answered question seems to help improve compliance. To ensure accurate interpretation of results, respondents to written surveys should also be told the significance of leaving an item unanswered. As will be shown in the following discussion of data extraction, a blank item can be very aggravating to a conscientious researcher.

The *layout* of a written survey is also important on paper and online. Experienced survey researchers know that an uncluttered, "user-friendly" form with ample white space (areas with no printing) promotes desired compliance and reduces undesired responses such as intentional sabotage and respondent fatigue. Bad layout can overcome good questions, so you have some reason to care about the visual appearance of survey instruments used to collect information for studies that might influence your decisions.

Finally, a survey's outcome can be influenced for better or for worse by the *language* used on a written form or by an interviewer. Words and contexts that are perfectly clear to health professionals who write the survey questions may have different or no meaning to the laypeople who answer them. A good survey will use the language of the respondents, not the researchers. Imagine, for example, the difference in a layman's (that is, not a health professional's) response to the two following questions:

- Has either of your parents suffered a cerebrovascular accident?
- Has either of your parents ever had a stroke?

A respondent without awareness of medical vocabulary may make up an answer to the first question rather than admit not knowing the definition of a cerebrovascular accident. Consequently, good surveys use the pretest stage to make sure that respondents understand the questions. Although survey forms are seldom included in research articles, significant questions are often presented in the discussion section, which gives you an opportunity to make your own assessment of the match between the vocabulary of the researchers and the vocabulary of the respondents. A study's results are cast in doubt if the surveyors and the respondents were not talking the same language.

Data Extraction

Having just addressed the importance of using common language, I am somewhat embarrassed to continue my review of survey-related issues with the use of a term I have "invented." (In other words, *data extraction* is not an established term within the world of statistics.) However, the absence of a professional term to encompass the underlying concern just might suggest a lack of proper attention to a problem that occurs much more often than most researchers would like to admit.

I use this made-up term, data extraction, to refer to the process of transferring the respondent's answers on a survey form to the digital data analyzed by the researcher. A study's results are distorted to the extent that something is gained or lost in the translation from survey form to computerized database because statistical analysis assumes that the data being analyzed are the actual research observations. Statistics does not compensate for errors in transcription. This is another potential problem that is difficult to evaluate based on the information typically provided in a research report, but it may be one worth pursuing if the published study gives you any hints that the researchers were less than careful in the process of extracting the data from the survey instrument.

The unreadable response is one of the biggest problems in this respect. Many people write on survey forms the way many doctors write on prescription pads: illegibly. If the researchers or their clerical assistants who enter data into the computer have to make guesses (for example, "Is that number a 2 or a 7?"), error can easily be introduced into the study. Consequently, giving people options to check can lead to better results than giving them blanks to fill in.

The ambiguous response is similarly vexing. Does a check halfway between the boxes marked "yes" and "no" mean "maybe?" Does it mean more "no" than "yes" if it is closer to "no," but still outside the box? (This problem reinforces the need to have clear instructions with good examples on every survey.) The problem can be bad enough if only one person is responsible for interpreting the responses when creating the database, but imagine how it is compounded if several people are reviewing the forms and entering the data—each with his or her own interpretation of ambiguous responses. Good research reports tell you how the problem was handled and what bias might have been introduced by the researcher's intervention.

Last, but not least, the blank response on a survey form may seem like a minor problem compared to the illegible or ambiguous response. Not so! Blanks create all sorts of problems in statistical analysis because they reduce the sample size on any given item. Computing the mean of a sample, the first step of just about all statistical tests, should be adjusted for blank responses. However, some researchers are rather sloppy in this regard, using the size of the whole sample rather than the number of subjects in the sample who actually responded on each separate item. The appropriate solution is to adjust the computation of each mean for the number of complete responses, but some researchers who do not want to sacrifice sample size actually estimate values for the missing items. Either way—failing to change computations to reflect the different sample size of each item or estimating (that is, making up) values for blank responses—a study's findings are biased.

My students and I know from experience that these "data extraction" problems occur fairly often. We become suspicious when a report of survey-based research fails to mention its approach to dealing with illegible, ambiguous, or missing responses. Our suspicions change to outright rejection of a study's findings when its authors use their own subjective judgments to fill in the blanks. Conversely, we have special respect for researchers who openly address problems with uninterpretable data and then leave well enough alone.

I could go on and on with a discussion of problems inherent in the ways data are collected, but that would defeat the purpose of this book. In the unlikely event that you wish to become an expert in this interesting area, a search of online scientific periodical indexes will allow you to locate articles that provide more detail on the problems I have summarized here and to identify additional problems not even mentioned in this chapter.

However, because you are a busy healthcare decision maker *and* because healthcare decision makers are intelligent people, I think I have provided enough information to make you aware of the problem. By taking the time to think about a particular statistical study in the spirit of the issues raised in this chapter, you will be able to spot serious data problems on your own.

Indeed, you have paid your dues by learning the basic scientific steps that must be taken before data are suitable for statistical analysis. You are ready to move on. So now, without further ado, the act you have been waiting for: the statistics!

Section III

The Different Types of Statistics

Always use the right tool for the job.

William C. Bauer, Sc.D.
My Father

5

Descriptive Statistics: The Foundation of Comparisons

Statistics is like a toolbox, full of different mathematical tools for working on numbers. Each tool in the box has its proper uses, but no single tool does everything. The tool you choose depends on the job. Like a skilled craftsperson who knows how to match the tool with the task, a good statistician knows which statistical tool is appropriate for the type of data to be worked. A decision maker who relies on the statistician's works needs to know enough to make sure that the job was done with the right tool.

Each of the final four chapters of this book describes a different type of tool—the statistical equivalents of wrenches, hammers, screwdrivers, and drills. (I hope you appreciate this metaphor, but please do not try to figure out which tool corresponds to which branch of statistics.) Much to my surprise, most of my students who have already taken one or two statistics courses have not been explicitly taught the differences in the types of statistical tools, so the four divisions of this last section reflect an important educational point. Each chapter title defines a meaningful division within the field.

Once a study has met all the scientific requirements related to experimentation and data, the first purely statistical task is to determine the desired analytical outcome so that the right statistical test (tool) can be selected for the job of data analysis. This step in the conduct of inquiry is very important. Many studies are seriously flawed by mismatches between the statistical tool, the type of data to be analyzed, and the desired framework of analysis. Bluntly stated, part of a decision maker's evaluation of critical information should address the match between the desired type of analysis (the job) and the type of statistical test (the tool) used to do it.

The vast majority of all statistical work reviewed by healthcare decision makers falls into one of four general categories. The categories encompass statistical tools that are designed to:

- Describe
- Compare
- Relate
- Explain and predict

The relevant question to be asked at the beginning of any statistical work is whether the end product is to be a description of the data from a single sample, a comparison of data from two or more samples, a study of the relationship between variables, or a model to explain the relationship between two or more variables in such a way that future values of the relationship can be predicted on the basis of understanding relationships from the past. The statistical toolbox contains different devices for getting each of the jobs done, depending on the materials to be transformed (for example, parametric or nonparametric data, small or large samples). The right tool can produce useful information for the decision maker. The wrong tool can produce useless or even harmful misinformation, which leads to bad decisions.

THE FIRST STEP: WHAT HAVE WE HERE?

Descriptive Analysis

No matter what the final product—a description, a comparison, a relationship, or an explanation and prediction—descriptive statistics should be the first tool out of the box because it is used to shape the data for the other products. Note that descriptive statistics produces very helpful information by itself. It is a tool worth using even when nothing more will be done with the numbers. But it is a tool that must be used if other products are desired because the two principal outputs of descriptive statistics—the measure of the middle of the numbers in a sample and the measure of the dispersion of the rest of the numbers around the middle—are required in the mathematical formulas for the other statistical tests. In other words, descriptive analysis must be performed in order to study comparisons, to identify relationships, and to make explanations and predictions.

Imagine a pile of numbers, figuratively speaking. This pile contains all the data gathered from each *case*, the researcher's terms for each subject in a sample. The *observations* (measured values of the variables on each case) will usually be entered into the database in the order in which they were collected, which means the data will initially appear in random order in studies where random case selection is required. As you imagine this pile of randomly distributed numbers, also think about the difficulty of trying to describe it in any way that would permit comparison with comparable data sets from other samples (which is what most of statistics is all about).

Since descriptive statistics is the foundation of the rest of statistics, I will develop a hypothetical database to illustrate the use of descriptive tools. As an author, I would plausibly be interested in the age of the people who buy my books, so I decide to do a study. I begin by randomly selecting 30 people who own copies of this book. (Ideally, I would draw a sample of 100 owners of the book because I prefer more powerful analysis, but $n = 30$ is theoretically acceptable and will not take up as much space as a sample of 100 cases.)

This simple-sounding first step in my hypothetical study immediately raises issues presented in the previous chapters of this book. For example, the method for identifying the population (that is, all purchasers of this book) would need to be evaluated if I want this to be a scientific study. I might end up with a bias toward younger owners if I sampled only students who bought the book for a statistics class. I would miss many older health professionals who are accustomed to learning on their own because they have less time available to take classes. So, for illustrative purposes, please join me in assuming that this book, like computer software, is registered to each owner. I can draw a random sample from the registration files and contact the 30 selected individuals.

Next, I need to determine the age of each person in the sample, a task that raises another research issue from the previous section. Should I ask for age, or year of birth? People are occasionally known to lie about their age, so year of birth is a more reliable response. Assuming that I get a 100 percent response rate and do not have to investigate the possibility of some systematic bias created by nonresponses, I end up with a preliminary data set that looks like Table 5.1.[1]

Well, what have we here? The values of the age variable appear in no particular order (consistent with the random selection of the 30 subjects), so the first step toward describing the sample's data is to put the values of the variable in order, as in Table 5.2.

Range

Now that the subjects' ages are arranged in order from lowest value to highest value, we can make the first descriptive statement about the data—defining the range of the distribution. The range is fixed by the endpoints of the distribution of the values in the sample. In this example, the range is 27 years to 82 years. The difference between the two endpoints—55 years—is another way to look at the range, but knowing the endpoints (that is, the lowest and highest values) is the most useful first step in descriptive statistics.

Note carefully that the case numbers are no longer in order once the values of the parametric variable, age, are ranked from lowest to highest. This should not cause any concern because the case number is not a variable in this study. Indeed, it is not a variable at all. It is nothing more than an identification tag that allows us to keep track of all the variables for each case. (We will add variables to this example in later chapters.) The case number is equivalent to the name and social security number on a patient record.

Because the case number is a name, it is sometimes called a *nominal variable*, but calling it a variable of any kind is misleading. Nominal variables have absolutely no quantitative significance. They measure nothing, so they should not be subjected to any statistical analysis. To my utter amazement, I occasionally encounter a study that performs statistical tests on a nominal variable. You need not read any further in a study if you encounter such silliness. Indeed, this practice may be the purest form of "garbage in, garbage out." (Imagine the irrelevance of the average Social Security number of persons included in a sample.)

[1] I have retained the data from the original edition of this book for these illustrations, which explains why the year is listed as 1995. Updating the year for each subsequent edition would be cumbersome, and it would quickly be outdated because new editions are not published each year.

TABLE 5.1

Age of Book Owners, March 1995

Case	Year of Birth (YOB)	Age (1995 minus YOB)
1	1958	37
2	1940	55
3	1961	34
4	1966	29
5	1944	51
6	1951	44
7	1959	36
8	1953	42
9	1932	63
10	1950	45
11	1938	57
12	1948	47
13	1964	31
14	1942	53
15	1947	48
16	1939	56
17	1936	59
18	1960	35
19	1958	37
20	1947	48
21	1913	82
22	1944	51
23	1946	49
24	1957	38
25	1968	27
26	1959	36
27	1948	47
28	1940	55
29	1947	48
30	1955	40

TABLE 5.2

Age of Book Owners, March 1995

Case	Year of Birth (YOB)	Age (1995 minus YOB)
25	1968	27
4	1966	29
13	1964	31
3	1961	34
18	1960	35
7	1959	36
26	1959	36
1	1958	37
19	1958	37
24	1957	38
30	1955	40
8	1953	42
6	1951	44
10	1950	45
27	1948	47
12	1948	47
15	1947	48
20	1947	48
29	1947	48
23	1946	49
5	1944	51
22	1944	51
14	1942	53
2	1940	55
28	1940	55
16	1939	56
11	1938	57
17	1936	59
9	1932	63
21	1913	82

Nominal variables have their place, just not in statistical tests. However, a closely related concept, the *ordinal variable*, is suitable for statistical analysis. Ordinal values indicate the numerical order of the cases once they have been ranked according to a true variable: 1 for the first on the list, 2 for the second, 3 for the third, and so on.

Ordinal variables are nonparametric. Why? Because the quantitative differences between the consecutively ranked cases are not necessarily equal. In the example I created for this book, the ordinal value of the youngest owner is 1, and the ordinal value of the next to youngest is 2—yielding a difference of 1 between the ordinal values. But the youngest owner is 27 years old, and the next owner on the list is 29 years old, for a difference of 2 years in the parametric variable, age. Ordinal variables lack the precision of parametric measurement, but they can yield useful information when worked with an appropriate nonparametric tool.

MEASURES OF CENTRAL TENDENCY

Describing and comparing piles of numbers requires a fixed frame of reference, the numeric equivalent of a surveyor's benchmark. The middle of a distribution is one of the two descriptive references embodied in the mathematical formulas of contemporary statistics. (The other descriptor, dispersion, is the subject of the next section of this chapter.) Indeed, the center of a sample's data is the point from which other statistical parameters are measured. It is "ground zero," the place where measurement starts.

This convention is logical, but it turns out not to be simple because the center can be defined in several different ways. The three common measures of central tendency—median, mode, and mean—are all important, but they are not necessarily identical. (The different concepts of central tendency are also called "first moments about the mean" in some statistics books.) Each of the three measures of central tendency and the differences between them should be memorized by anyone who reads published studies because authors will use the terms without defining them. (Authors will also use them incorrectly on occasion. You might be bamboozled by their ignorance if you cannot immediately recall the important distinctions between the measures of central tendency.)

Median

The median is the middle value, the halfway point when the data from all cases are laid out in a line. My favorite device for remembering the meaning of the median is to think of a four-lane divided highway. The median is the middle of the road; it divides the highway in half. The statistical median is the number at the middle of the ordered values.

To find the median, put all the numbers in a distribution in an ordered row. Then, find the number that is at the middle of the distribution. Here are all the values from Table 5.2 of my hypothetical database laid out in order:

> 27, 29, 31, 34, 35, 36, 36, 37, 37, 38, 40, 42, 44, 45, 47,
> 47, 48, 48, 48, 49, 51, 51, 53, 55, 55, 56, 57, 59, 63, 82

The sample size is 30, so the midpoint occurs between the 15th and 16th numbers in the sequence. Since value of number 15 and value of number 16 are both 47, the median for this distribution is 47.

A minor problem can occur in situations like this, where the sample size contains an even number of values, if the two values that bracket the middle are not the same. Common practice is to define the median as the midpoint between two unequal values that border the middle (for example, 47.5 if the 15th value had been 47 and the 16th had been 48), but this is a little misleading since 47.5 was not an actual value. Be aware of the potential problem, but do not lose any sleep over it. The problem does not exist when the sample size includes an odd number of values because a true middle value exists in that case.

Mode

The mode is the value that occurs most frequently in a distribution. Mode comes from a French phrase commonly used in English—*à la mode*—which means stylish, the most popular fashion, the thing seen most often. (It does not mean that the numbers are covered with ice cream, but the practice of topping off a piece of pie *à la mode* with ice cream comes from the fact that it was once the rage.)

To find the mode, look at frequency (that is, the count) of each value in a distribution to find the value that occurs the most. Creating a *frequency distribution* is the first step in the process, and one you would be wise to learn to do because frequency distributions can help you understand the databases you create in your own work as a healthcare decision maker. Table 5.3 is the frequency distribution for my hypothetical database of 30 randomly selected owners of this book.

The mode is 48 years because 48 occurs more frequently than any other value in this particular pile of numbers—more elegantly called a *distribution*. Distributions can also be multimodal. For example, if another value had also occurred three times, the distribution would be bimodal.

TABLE 5.3

Frequency Distribution: Age of Book Owners, March 1995

Value	Frequency
27	1
29	1
31	1
34	1
35	1
36	2
37	2
38	1
40	1
42	1
44	1
45	1
47	2
48	3
49	1
51	2
53	1
55	2
56	1
57	1
59	1
63	1
82	1

The frequency distribution is an important first step in organizing and describing data. However, it can be rather cumbersome when many of the values occur infrequently, as in this example where most of the numbers occur only once. To simplify matters without losing important information about the distribution, the data are divided into logical categories of equal width. Each division is called a *cohort* or *cell* and includes the total count of all values in the cell. In practice, open-ended cells are permitted at the top and bottom of the distribution. Consistent with the common epidemiological practice of classifying age in 10-year cohorts, the frequency distribution for my study can be meaningfully simplified as illustrated in Table 5.4.

The final step in developing this particular measure of central tendency is to present the data in a chart called a histogram. The histogram for my hypothetical data set is illustrated in Figure 5.1.

The key to creating informative histograms is knowing how to determine a sensible cell width. I know of no uniformly valid rule to guide the task, other than that there is no substitute for experience. Since histograms are extremely useful for presenting data, I suggest you learn how to make them by practicing with your own data. Just about any spreadsheet or statistics program will allow you to make very nice-looking histograms on your own Mac or PC.

Graphic presentations of statistical information should be clearly and adequately labeled with a title and additional footnotes as necessary. A brief description of the data and the date of the data should appear as a title above the table or graph. Footnotes should provide additional details of interest to the expected users of the information, such as the source

TABLE 5.4

Frequency Distribution by Cohort: Age of Book Owners, March 1995

Value	Frequency
Under 26	0
26–35	5
36–45	9
46–55	11
56–65	4
Over 65	1

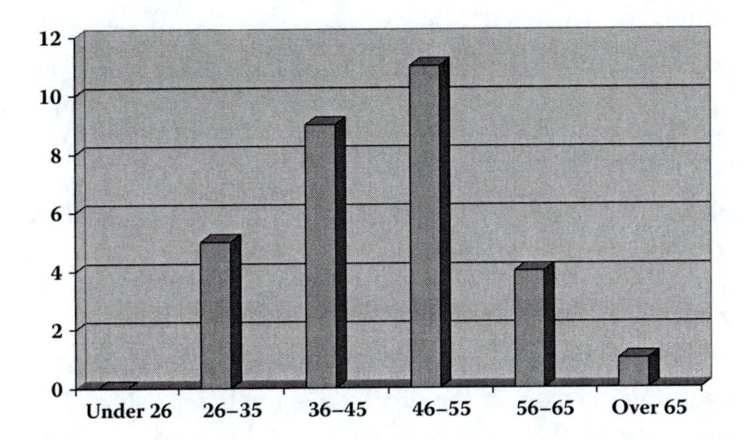

FIGURE 5.1
Frequency distribution by cohort: Age of book owners, March 1995.

of the data (from the American Medical Association's *Survey of Medical Practice* or the Internal Revenue Service, for example) and detailed definitions of the data (such as annual gross billings to third-party carriers, or personal income to physicians as reported on Form 1040).

Sloppy or incomplete labeling is a real problem in published reports of healthcare studies. For example, the date of data collection is often missing, which can be aggravating because knowing the time period of data is very important if data from different studies are to be meaningfully compared. And even when the date is reported, the reader cannot always tell whether the date refers to the time when the data were collected or the time when the data were reported. The time lag between data collection and publication is commonly 1 or 2 years in healthcare studies, so an article's failure to be clear on this point can make the data fairly useless in a fast-changing, highly competitive market such as healthcare. Online publication can be much faster, but it does not eliminate the importance of labeling data with the time they were collected.

Mean

The *mean* is the principal measure of central tendency. It is usually the first factor to be computed in statistical equations and is one of the two summary statistics, along with the standard deviation, that provides the mathematical basis for comparing two or more distributions. It is not necessarily better than the other measures of central tendency in any absolute sense, but it has one significant advantage over its two counterparts. The mean can be manipulated algebraically; the median and the mode cannot. Since algebra was the main mathematical tool used when statistical concepts were developed during the nineteenth century, the mean became the building block of statistical analysis. (Do not waste any time worrying about this historical fact because I believe it is on the way to becoming irrelevant, as argued in the introduction, "Slide-Rule Science in the Computer Age?")

To compute the mean value of a variable, add together all the values of the variable and then divide the sum by the number of values. This operation is expressed by the standard equation for the mean:

$$\bar{x} = \frac{\sum_{i=1}^{n} x_i}{n} \tag{5.1}$$

The large, angular shaped letter is sigma, the standard mathematical symbol that instructs you to sum a series of numbers. The little i just below the sigma tells you where to start the series, so $i = 1$ means start with the first number, or 27 years in my hypothetical data set. The n on top of the sigma tells you to add all the numbers in the sample since it is not limited, so add all the values up to and including 82 years. (If the notation on top read $n = 15$, you would know to add only the first 15 numbers in the set.) The n below the line tells you to divide the sum by the sample size. All this equals "x-bar," which is the mathematical nickname for the mean, which is also known as the arithmetic average. If you wish to make the computation for yourself, you should get the value of 46.0 years as the mean age of owners of this book.

Although the mean is mathematically convenient, it has one potential problem that is apparent in my hypothetical example: it can be substantially influenced by extreme values. The 82-year-old owner of this book really brings up the average, in spite of the cluster of owners in the two lowest cohorts (26–35 and 36–45). The median (47 years) and the mode (48 years) give different values of the center of the distribution, but the mean (46.0) will serve as the measure of central tendency for subsequent statistical analysis. This outcome shows the importance of having three measures of central tendency and knowing the difference between them. It also shows the importance of looking at the complete distribution of the data so you will be aware of any *outliers*—extreme values that influence the mean.

If you are thinking that a simple solution would be to remove the octogenarian from the sample, you are forgetting the earlier discussion of random selection (Chapter 4, Sample Size). Statistics relies on random selection to prevent bias, so removing the outlier would be a violation of the rules on which statistical analysis is built. If we are unhappy with the distortion created by including the 82-year-old owner in the random sample, the scientifically appropriate solution is to draw a larger random sample, not to inject human judgment by removing the subject.

I occasionally encounter published studies that remove an outlier from the sample without thinking twice about it, so be on your guard. (Indeed, deleting both the high and low values from a sample before computing the mean used to be a common practice.) Random selection is not always perfect, but the discussion at the beginning of the next chapter will show why it is a central foundation of statistical analysis. Intervening to "correct" perceived imperfection in a random sample may seem logical, but it is a gross violation of a theoretical assumption that makes statistics possible.

MEASURES OF DISPERSION

Overview

Defining the middle of a distribution is a sensible step in the process of turning data into information. Median, mode, and mean are good summary statistics. They help us understand the numbers in a sample. However, by themselves, the commonly accepted measures of central tendency do not provide enough information to accomplish the principal task of statistics: the study of differences within or between samples. Defining the dispersion of numbers within a distribution is just as important as defining the middle.

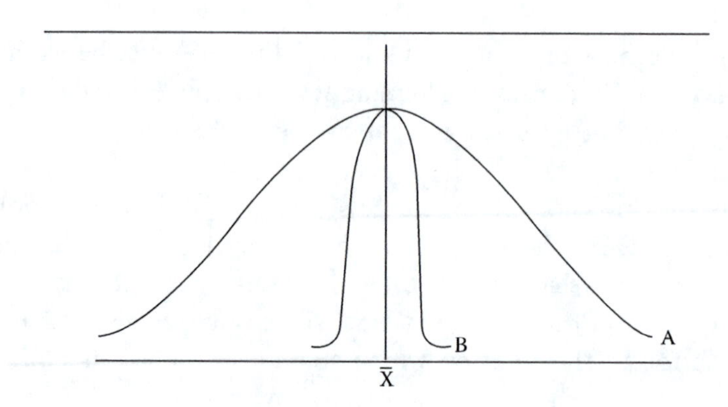

FIGURE 5.2
A demonstration of differences in dispersion.

Figure 5.2 demonstrates the theoretical importance of dispersion. Sample A and Sample B have the same mean, but the values in Sample A are widely dispersed in comparison to those in tightly clustered Sample B.

We might assume that the samples were similar if the mean were the only descriptive statistic available to us. However, Figure 5.2 shows how wrong we could be if we used only a measure of central tendency to compare distributions. The dispersion of values around the center is every bit as important as the center itself.

As will be shown in Chapter 6, statistical equations incorporate quantitative expressions of both the middle and the spread. Do not let your decisions be swayed by a study that does not present both. In particular, resist the temptation to be influenced by media reports that mention only one of the descriptive parameters. Journalists are known to create misleading stories by reporting only the measure of central tendency in stories about research. In particular, they often confuse the median and the mean and treat these different concepts as synonyms for average. The concept of average is essentially meaningless in the popular press.

Fortunately, measures of dispersion—often categorized as second moments about the mean—tend to be used more consistently, when they are used at all. The simplest measure of dispersion in common use is the range, which was discussed previously in this chapter (see "Range") since it is immediately defined when data are put in order. Several other measures of dispersion have been developed over the years, but only one of them needs to be covered in this book because it is the only one likely to be used in studies that will be reviewed by healthcare decision makers.

(In other words, I am sparing you a lengthy discussion of information that is interesting in theory but useless in practice. You can find in-depth explanations of the infrequently used measures of dispersion in a typical statistics text if you have the time and interest.)

Standard Deviation

The standard deviation is the commonly accepted approach to expressing the dispersion of values within a set of data. Conceptually, the standard deviation is a measure of the average difference between each of the variable's values in a distribution and the mean of all the values. The more the values are spread out, the higher the standard deviation should be. However, computing the standard deviation is not as simple as it sounds because the sum of differences between the mean and each value of the variable is zero, and the computational approach needs to give appropriate weight to outliers (such as the 82-year-old in the hypothetical sample of owners of this book).

The mathematical equation for the standard deviation of a sample is a bit intimidating in print:

$$S = \sqrt{\sum \frac{(X - Xi)^2}{n - 1}} \tag{5.2}$$

However, the computation is actually pretty simple. The first step is to calculate the difference between the mean and each of the values in the sample. The result of this computation, the mean difference for each of the 30 cases in my hypothetical data set, is shown in the third column of Table 5.5.

Note that the sum of the mean differences is zero, so the average mean difference would also be zero ($0 \div 30 = 0$). This produces a useless measure of dispersion because anyone can plainly see that the numbers are spread out. Something more needs to be done. The conventional solution is to square the mean differences from the sample and then sum them, which produces a value of 3,882 as shown in the fourth column of Table 5.5.

The next step in computing the standard deviation for a sample is to divide the numerator, the sum of squared deviations, by the sample size minus one (the $n - 1$ in the denominator of the equation), which equals

TABLE 5.5

Initial Steps in the Computation of the Standard Deviation

Case	Age	$\bar{X} - X_i$	$(\bar{X} - X_i)^2$
1	37	9	81
2	55	-9	81
3	34	12	144
4	29	17	289
5	51	-5	25
6	44	2	4
7	36	10	100
8	42	4	16
9	63	-17	289
10	45	1	1
11	57	-11	121
12	47	-1	1
13	31	15	225
14	53	-7	49
15	48	-2	4
16	56	-10	100
17	59	-13	169
18	35	11	121
19	37	9	81
20	48	-2	4
21	82	-36	1,296
22	51	-5	25
23	49	-3	9
24	38	8	64
25	27	19	361
26	36	10	100
27	47	-1	1
28	55	-9	81
29	48	-2	4
30	40	6	36
	SUM =	0	3,882

29 in the example since the sample size is 30. This step yields a quotient of 133.86. The final step is to take the square root of 133.86, essentially undoing the squared function that was introduced in the numerator to overcome the problem of the mean differences adding up to zero. This computation produces a value of 11.57 years, the standard deviation in the age of the sample of persons who own this book.

The denominator of the equation for the standard deviation merits two comments:

- First, the term (n – 1) defines *degrees of freedom* in the analysis. This arcane concept reflects the fact that the number of values free to vary is reduced by one once the sample mean has been calculated. Although the concept of degrees of freedom has no direct impact on a decision maker's ability to understand statistical analysis in research reports, it is important in the selection of decision values from statistical tables—in the unlikely event that you ever need to consult a table to find the threshold value of a statistical test at which the null hypothesis would be rejected. (This task is now handled quite effectively by software and computers. Most of us will live the rest of our lives without ever having to consult a statistical table that incorporates sample size and degrees of freedom.)

- Second, the impact of adjusting the formula for degrees of freedom—that is, of dividing the numerator by (n – 1)—decreases as sample size increases. The larger the denominator, the smaller the quotient and the final value of the standard deviation. For example, all other things being equal, $n = 30$ produces a smaller value for the standard deviation than $n = 10$. The statistical significance of tests for differences is strengthened by smaller values of the standard deviation, which is another reason why bigger is better when it comes to the size of the random sample.

The standard deviation has no absolute meaning in any cosmic sense; it cannot be reduced to a demonstrable, repeatable truth. It is an artificial construct developed back in the early nineteenth century by a mathematician who was devising simple ways to summarize a lot of astronomical numbers. Other approaches to expressing variation have also been

ESTIMATORS

Statistics has at least three common meanings, a fact that can create confusion. Generically, the term refers to a specific form of quantitative analysis, such as the subject of this book. In lay usage, statistics is used as a synonym for data, as in "Let's collect some statistics." To statisticians, the term encompasses the summary measures used to describe and compare numbers. For example, the various measures of central tendency and dispersion are statistics. These moments about the mean and the other products of statistics' many equations are also known within the field as *estimators.*

Nonstatisticians who use statistical information should be aware of the concept of estimation because it reflects the imprecision inherent in statistics. In one sense, the very existence of three different measures of central tendency—median, mode, and mean—demonstrates the lack of a single, precise, universally accepted way to define the middle of the data from a sample. Likewise, the standard deviation is just one estimate of dispersion. From the broader perspective of sampling and probability, summary statistics calculated with data from randomly drawn samples are estimates of true population parameters. If we had the time and the money to study entire populations, we would not need statistics. The declining costs and rising power of computers, along with the digital transformation of our data bases, will surely reduce the need for statistics. We will not need to make inferences from samples because we will increasingly have complete population data at our fingertips.

The history of statistics is largely the story of mathematicians and other scientists developing new and hopefully better ways to estimate population values from sample data. The perfect estimator has not yet been developed (and probably never will be), so anyone who relies on statistical analysis should keep in mind that statistics is still essentially an art dedicated to making the best possible guess under the circumstances. Statisticians are being honest, not cute, when they use the term *guesstimate.*

developed, but none has gained the same level of acceptance. The standard deviation does provide us with a useful benchmark for comparing the dispersions of different distributions.

The important point is that the traditional approach to looking for differences between samples requires having a consistent way to measure dispersion from sample to sample, and the standard deviation serves the purpose well when quantitative analysis is reduced to comparisons of descriptive statistics. It is not very sophisticated in comparison with new analytical techniques made possible by the power of modern computers, but it has been widely accepted and universally understood for well over 100 years. In other words, it is a tradition—a paradigm waiting to be displaced.

6

Inferential Statistics: Studies of Differences

Shaping sample data with descriptive tools is only the first step in statistical analysis. Several useful and interesting tasks can be performed with other techniques in the statistical toolbox. The most common of these tasks is undoubtedly the comparison of data from different samples to assess the likelihood that the subjects in the samples come from different populations. The tool to study differences is *inferential statistics*. Literally, its name reflects the need to make inferences—to arrive at a conclusion by reasoning from incomplete evidence, according to *Webster's*—since we do not know everything about the situation. (This sounds like the daily circumstance of today's health professionals, yes?)

The proper application and the meaningful interpretation of inferential statistics both require understanding the tool's theoretical basis. Researchers who do not know the underlying theory can create misleading information, and decision makers who do not understand the theory can be misled. I have concluded after a quarter century of teaching that the essential theoretical foundations of inferential statistics are not adequately presented in most statistics texts or courses. Therefore, consistent with my goal of emphasizing understanding over believing, this chapter puts the emphasis on theory in order to correct the imbalance that leads to bad decisions. (My students are almost all very smart, so I feel comfortable placing the blame on the instructional side. If the theory behind inferential statistics had been well presented, they would have learned it.)

THE NORMAL DISTRIBUTION

Inferential statistics is made possible by the characteristics of a very special distribution, the famous "bell-shaped" curve.[1] It is also called the normal or Gaussian distribution (after a nineteenth-century German mathematician, Carl Gauss, who elaborated the properties of a distribution originally developed during the seventeenth century by a Swiss theologian named Jacques Bernoulli).[2] It is illustrated in Figure 6.1.

The normal distribution has three properties that allow us to make probabilistic statements about differences:

1. The three measures of central tendency are identical in the normal distribution. Median, mode, and mean all have the same value.
2. The curve is perfectly symmetrical (and looks like a bell). The distribution of values above the middle is a mirror image of the distribution of values below it.
3. The shape of the normal distribution can be expressed in a mathematical equation that allows us to define the area under every part of the curve.

This third property is especially useful. By definition, we know that half the values under the curve are below the midpoint (that is, the median, mode, and mean), and half the values are above it. More to the point of statistical analysis, however, differential calculus can be used to compute additional breakdowns in the area under the curve. (I will bet that you never expected to see the day when differential calculus would prove useful!) The area defined by the mean plus one standard deviation includes 34 percent of the values in the distribution (the same as 34 percent of the area under the curve). Because the normal distribution is perfectly

[1] The distribution gained considerable attention through the controversial book by Richard Herrnstein and Charles Murray, *The Bell Curve: The Reshaping of American Life by Differences in Intelligence* (New York: Free Press, 1994). I am pleased to observe that the heated debate surrounding their analysis centered on the fundamental issues of science and data that were presented in the first two parts of this book. I wish commentators had been equally concerned with scientific and quantitative integrity during the 1993–94 debate over healthcare reform. I think we would have saved a lot of time and money if they had bothered to evaluate the debate from these important perspectives.

[2] Jean-Jacques Droesbeke and Philippe Tassi, *Histoire de la Statistique* (Paris: Presses Universitaires de France, 1990).

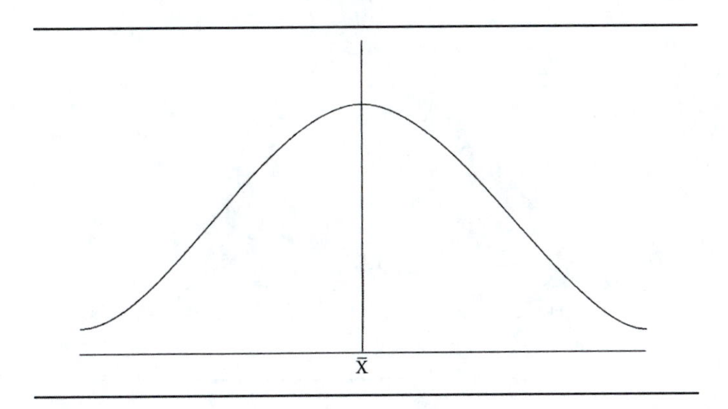

FIGURE 6.1
Normal distribution.

symmetrical, the area defined by the mean minus one standard deviation also includes 34 percent of the area under the curve. Add these two facts together, and we know that 68 percent of the values in the distribution lie between one standard deviation below the mean and one standard deviation above it.

Next, differentiating the equation for the standard normal curve shows that the mean plus two standard deviations includes 48 percent of the distribution. By extension, 96 percent of the area under the curve—the same as 96 percent of the values in the distribution—lies between two standard deviations below the mean and two standard deviations above it. Finally, the mean plus and minus three standard deviations includes 99 percent of the area, or 99 percent of the values in the distribution. Please note that these neat relationships hold only for the normal distribution, the bell-shaped curve. They are illustrated in Figure 6.2.

One slight modification in these relationships is needed to adjust the relative portions of the underlying distribution for standard statistical practice. Researchers traditionally like to use a 95 percent figure when applying the normal distribution to their statistical analysis, and 95 percent of the area under the curve is encompassed by the region between −1.96 standard deviations below the mean and +1.96 standard deviations above the mean. The area defines the *95 percent confidence interval.* (This distinction would not be necessary if scientists had been inclined to be 96 percent confident in their analyses.) The 95 percent relationship is very important, so Figure 6.3 provides an illustration to imprint the concept in

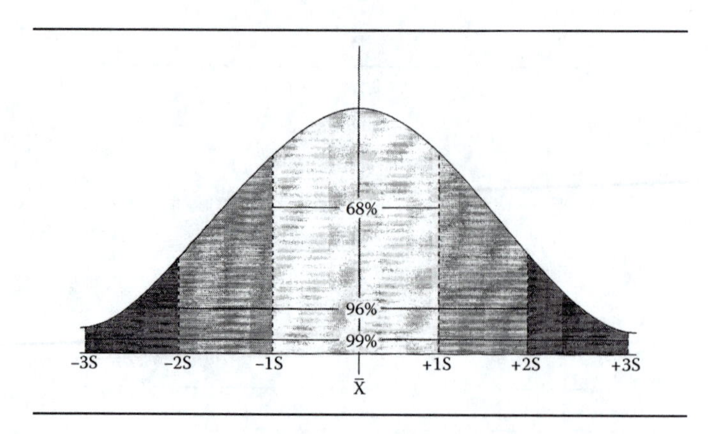

FIGURE 6.2
Relative interval areas under the curve: Normal distribution.

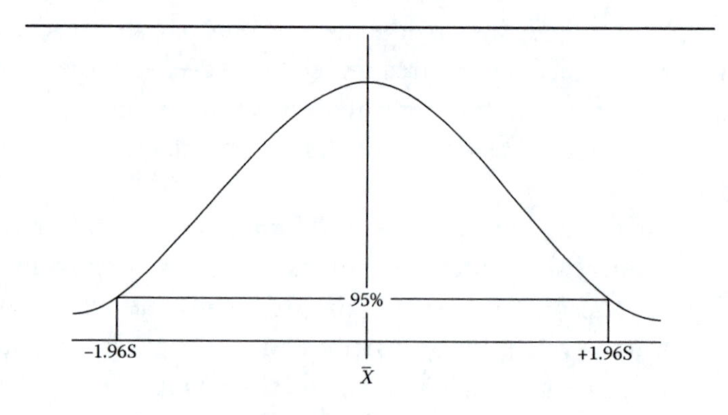

FIGURE 6.3
Ninety-five percent interval: Normal distribution.

your mind and give you a visual framework for the subsequent discussion of inferential statistics.

Inferential statistics is made possible by an obvious relationship between the area under the normal distribution's curve and the probabilities associated with drawing values randomly from the distribution it represents. To understand this very important theoretical point, imagine a bell-shaped curve representing the distribution of a sample with 100 cases (n = 100). Any one of the 100 cases has an equal probability of being picked, but a case with the modal value has the highest probability of being picked because the mode occurs most frequently, by definition.

Likewise, a case with the mean value is most likely to be picked because the mean and the mode are identical in a normal distribution, and the chances of drawing a case below the median are equal to the chances of drawing a value from above the median because the median divides the values in half. Last, but not least (in other words, this point is important), a value between −1.96 standard deviations and +1.96 standard deviations will be drawn 95 times out of 100. Cool!

Just in case you are wondering where statisticians found this useful relationship, here is a very brief digression on the origin of distributions. (If you do not care, skip two paragraphs. This material will probably not be on the test.) The bell-shaped curve is just one of many distributions that embody the probability of important events and help us conduct statistical analysis. The normal distribution occurs commonly in real life. Height, weight, test scores (once upon a time, before grade inflation), and many biological and behavioral variables are found to be normally distributed. The normal distribution corresponds quite nicely with the types of common events that end up being reported in healthcare journals.

However, the normal distribution does not cover all situations. Consider an event that occurs rarely, such as randomly drawing the single green ball from a box that also contains 999 otherwise identical orange balls. The probability function for this one-in-a-thousand event can be worked out theoretically and verified empirically. It produces the *Poisson distribution*, which does not look anything like a bell-shaped curve. It is very lopsided and skinny. Indeed, a unique distribution can be calculated for every possible combination of green and orange balls in a box, and non-normal distributions such as the Poisson are relevant in some situations. Do not hesitate to ask a statistician for an explanation and help in the infrequent situation where you encounter one of these other distributions.

Back to the topic at hand, let us illustrate the relationship between distribution of values and probability with the hypothetical example of people who own this book. Each and every one of the 30 owners in the sample would have an equal probability of being selected in a random drawing, but a 48-year-old has the highest probability of being drawn because 48 years is the most common age in the sample (3 out of 30). Owners who are 36, 37, 47, or 55 are next most likely to be selected because the sample contains two owners with each of these ages.

Further, an owner older than the median (that is, 47 years, the age that divides the group of owners in half) is just as likely to be drawn as an

PROBABILITY AND ODDS

By definition, the sum of the probabilities of all possible outcomes of an experiment equals 1.00. If one outcome is a sure thing (that is, no other outcome is possible), its probability is 1.00. At the other extreme, something that absolutely cannot happen has a probability of 0.00. If an event has only two possible outcomes and they are equally probable, the probability of each equals 0.50. For example, the probability of getting a head on each toss of a fair coin is 0.50; it is written as $p = 0.50$.

Simple probabilities are easy to calculate when you know the total number of possible outcomes and the number of times that each individual value could occur. For example, if a box contains 100 tennis balls, and 22 of them are green, the probability of randomly drawing a green ball is 22/100, or $p = 0.22$. Assuming all the other tennis balls are orange, the probability of drawing an orange ball at random is 78/100, or $p = 0.78$. The basic concept of probability is sufficient for the purpose of this book, but many complex applications of probability theory have been developed. Consult a textbook on probability if you are interested in learning more.

The concept of odds is a bit different. Take, for example, the flip of a fair coin. Since heads and tails are equally likely, the odds are 50:50, which yields an odds ratio (50 divided by 50) of 1.00 which is not the same as $p = 1.00$. If the odds of some event (for example, the estimated chances of Democrats or Republicans winning an election) are 60:40, the odds ratio is 1.50 (60 divided by 40). Statistics deals with probabilities, so be careful not to confuse probabilities with odds.

owner younger than 47. (By the way, the probability of drawing someone younger than 47 is not $p = 0.50$ because 2 of the 30 owners are 47 years old. The probability of drawing someone from below the median is 14/30, or $p = 0.4666$, because only 14 owners are below the median. The same probabilistic relationship applies to drawing an owner with an age above the median.)

However, the probability of drawing the mean value in my hypothetical example is not equal to the probability of drawing the median value or the modal value—a key feature of normally distributed data. Why not? First, the mean age is 46 years, yet no 46-year-old owners appeared in my randomly selected sample of 30 cases. The mean value cannot be drawn because it is not there. Second, the mean is not equal to the mode (48 years) or the median (47 years). Therefore, by definition and by observation, we must conclude that my hypothetical sample is not normally distributed. In fact, if you look carefully at the random samples used in studies intended to influence healthcare decision makers, you will quickly discover that samples are almost never normally distributed. Bell-shaped curves with special attributes such as 95 percent confidence intervals are remarkably rare in research studies.

We seem to have a problem here. I have already shown that statistics provides tools for comparing randomly drawn samples including at least 30 cases and that these probability-based comparisons are made possible by the special characteristics of normally distributed data. Have I wasted your time by developing an example that is unusable? Do researchers keep drawing samples until their sampling produces normal distribution? Please bear with me while I explain how statistics handles this problem. What follows is the theoretical foundation of modern statistics. I hope you will understand it so you can decide whether to believe it.

THE CENTRAL LIMIT THEOREM

Since the founders of statistics were primarily empiricists—government officials and scientists in search of reasonably efficient estimators to simplify the task of understanding databases that were too large to be comprehended by themselves—they discovered a special property of random selection of data from distributions that were not necessarily normal. Even when a population is non-normal, a special procedure can be used to produce a normal distribution. Here is how it works.

The process requires drawing repeated small samples, usually $n = 10$, from a known population that is not normally distributed. The mean of each sample is calculated and recorded. An amazing thing happens when the distribution of these sample means is plotted as a histogram. *As the number of samples grows larger, the mean of the distribution of sample means converges on the true mean of the population.* The larger the number of samples, the more closely the mean of the sample means estimates the true population mean. Indeed, as the number of samples approaches infinity, the mean of the sample means becomes the population mean. *Further, the distribution of the sample means is a bell-shaped curve.* (I can assure you that the procedure does work. Like many graduate students taking statistics, I once spent many hours doing this exercise, and it produced exactly the results promised by the theory. If you need to be convinced but do not want to take the time, hire your kids to do it.)

Voilà! We have a demonstrably effective method for creating a normal distribution from data that are not normally distributed. The fact that research samples are almost never normally distributed does not prevent us from making probabilistic inferences based on the special characteristics of a normal distribution because we can create a normal distribution that converges on the actual value of the population mean by applying the central limit theorem. Pretty neat, theoretically speaking.

STANDARD ERROR (OF THE MEAN)

Thanks to the central limit theorem, we know we can come very close to determining the mean of a population by drawing a very large number of small samples, plotting the distribution of these sample means, and calculating the mean of the sample means. But there was one problem with putting this theory into practicing back when statistics was developed. Drawing all those samples and computing all those sample means and the mean of the sample means took a lot of time and effort, and statistics was developed to save time and effort. (Remember that the slide rule was the state-of-the-art computational aid at the time. Mechanical adding machines were several decades in the future, and the electronic calculator would not arrive for almost a century.)

Consequently, the mathematically sophisticated but computationally challenged founders of statistics developed an estimator—a shortcut, if you will—to approximate the mean of the sample means. They explored the theoretical trade-off between using a lot of small samples (for example,

$n = 10$, as used in the development of the central limit theorem) and using one bigger sample to see just how large a single sample would have to be to produce an acceptable approximation of the population mean. Their efforts to simplify the process of finding a population mean produced the concept of *standard error.* The standard error is used in comparative statistics because it embodies the likelihood of a difference between a sample mean and a population mean; the standard deviation by itself does not.

The standard error of the mean is a straightforward application of the general concept of standard *error.* Its very name, standard error, reflects the fact that the mean of a single sample is an imperfect approximation of the population mean, but the theory and mathematics behind the concept suggest that a randomly drawn sample of 30 cases can provide an acceptable approximation of a population mean—one that can be used for purposes of comparing different samples to assess the likelihood that they came from the same population.

Note carefully the use of *acceptable* in the previous sentence. A sample size of 30 is the minimum that produces 95 percent confidence when a single random sample is used to approximate a whole population. (As I argued when discussing sample sizes back in Chapter 4, I feel much more comfortable with larger samples.) The equation for the standard error of the mean shows why:

$$SE_{\bar{x}} = \frac{s}{\sqrt{n}} \tag{6.1}$$

The equation says that the standard error of the mean is the standard deviation for a sample divided by the square root of the sample size. Since we would like the error to be as small as possible—that is, we want the estimated mean to be as close to the population mean as possible—a larger denominator is better than a smaller denominator. Obviously, the way to increase the value of the denominator is to increase the sample size. The larger the sample size, the smaller the estimated error for a given standard deviation. The probability of error still exists as the sample size increases, but at least it declines.

QUICK REVIEW

You have now seen the theoretical concepts that lead up to inferential statistics, the branch of statistics used in most studies that are intended to influence the decisions of health professionals.

- A sample of 30 or more cases is drawn randomly from a population. Descriptive statistics is used to define the middle (mean) and the spread (standard deviation) of the distribution of values in this sample.
- Applying the theory behind the central limit theorem, the standard error of the mean is then computed with a formula that adjusts for the possibility of error inherent in estimating the population mean from a sample. The standard deviation for the sample and the standard error of the mean provide the basis for the most common statistical comparisons of different samples.
- When all these preliminary steps are done properly, the summary statistics (estimators) of different samples can be compared through the use of various mathematical equations. The result is a tool for testing the null hypothesis (H_0) that the different samples came from the same population with a specified level of confidence.

Behind all of these steps is the assumption of randomness. If the sample selection process is not random, the statistical analysis is flawed, period. *The theory of statistics requires randomness*, so pay special attention to sampling techniques when you evaluate research-based studies. Do not base decisions on studies that relied on anything other than the luck of the draw to collect the data.

HYPOTHESIS TESTING

Now for the fun part: using statistics to analyze research data. Assuming that a study was conducted according to the principles of scientific research, as covered in Section I, and that it has paid appropriate attention to the quality of data, as covered in Section II, statistics has some special tools for evaluating the quantitative information. Section III provides a conceptual understanding of the basic tools for analyzing research data to see if something makes a difference.

Please note that I just promised to explain the *basic* tools. Statisticians have developed many different versions of these tools to meet the needs of special situations. Explaining all the tools and their nuances would defeat the purpose of this book, making it every bit as cumbersome as the statistics text that probably baffled you when you previously took the course. I am comfortable, however, that this book explains the general theory of statistics well enough to help you understand the specialized statistical tools, if and when you encounter them.

That said, I have concocted a hypothetical experiment to illustrate how inferential statistics actually works. I am interested in knowing whether this rather unconventional book makes any difference in a typical health professional's understanding of statistics, so I will start this exercise by specifying a null hypothesis.

H_0: Reading *Statistical Analysis for Decision Makers in Healthcare* does not affect a health professional's understanding of statistics.

In other words, I am going to conduct a study to test the proposition that this book you are reading simply does not make a significant difference, that it does not affect your understanding of statistics one way or the other. The acceptance or rejection of this hypothesis should help you make an important decision whether or not to buy and read this book. If the null hypothesis is accepted, you might as well spend your time and money in a more productive way because the book does not make a statistically significant difference.

The hypothesis leads to an experiment, beginning with the random selection of 60 students from the entering class at the University for Health Professionals. Next, the researcher randomly assigns these students to two statistics classes so that $n = 30$ in each sample, and the same standardized statistics test is administered to both classes to establish baseline scores before the students are exposed to any formal instruction in the subject. Each class is then taught the same way by the same statistics professor using the same text, *Statistics the Traditional Way*. However, we introduce an experimental effect by additionally assigning this book, *Statistical Analysis for Decision Makers in Healthcare*, to one of the classes. At the end of the semester, the students in both classes take an identical final examination (not the test used to establish the baseline).

As a result of the experiment, we can produce summary statistics from two distributions of data:

1. The mean and standard deviation of the test scores of the statistics students in the class that only used *Statistics the Traditional Way* as the assigned text (the control group).
2. The mean and standard deviation of the test scores of the students who also used *Statistical Analysis for Decision Makers in Healthcare* (the experimental group).

The first step in the basic statistical procedure for making a probabilistic inference about differences between two samples is to compare the mean of both samples. If using this book probably did not make a significant difference, as posited by the null hypothesis, we would expect a small difference between the sample means, as shown in Figure 6.4.

On the other hand, as shown in Figure 6.5, we would expect to see a big difference between the sample means if using this book made a big difference in students' understanding of statistics. (This approach to measurement raises the question whether the test score is a valid measure of students' understanding. I will let you think through that issue on your own.)

The first question of inferential statistics is how much difference can exist between the two means before they are likely to have come from different populations—specifically, that the students who used this book in addition to the traditional text probably understand statistics differently from the students who did not use it, with everything else being the same (controlled). The closer the sample means, the more likely they came from

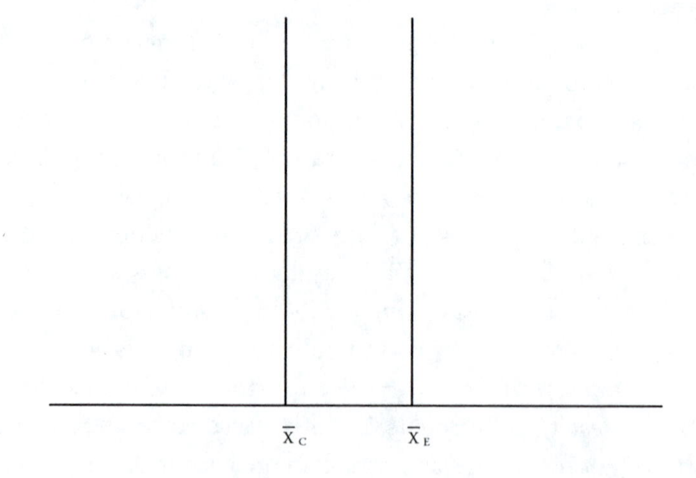

FIGURE 6.4
Small difference between sample means.

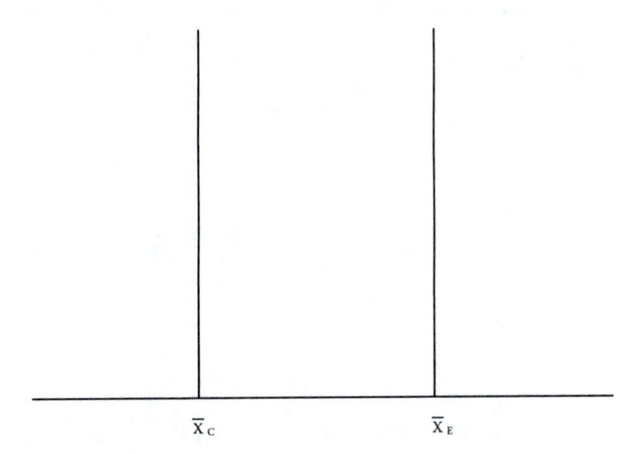

FIGURE 6.5
Large difference between sample means.

the same population—which means that the experimental effect did not make a difference. This initial concern with the difference between the sample means is reflected in the numerator of the equations for the basic tools of inferential statistics.

However, we have also seen that dispersion is an equally important factor in defining a set of data, so standard deviations—converted to standard errors in order to bridge the gap between the practical reality of a single sample and the central limit theorem's expectation of many samples—also need to be brought into the picture when two or more samples are compared. The denominators of the equations of inferential statistics effectively combine the variation within the samples by "pooling" the standard errors of each distribution. Before looking at the equations, let us take one more look at the comparison of means, with dispersion added to the picture.

The two means pictured in Figure 6.4 look close together in comparison with the two means pictured in Figure 6.5, but the addition of dispersion (variance) can give a totally different picture. The two sample means in Figure 6.6 are just as close to each other as the two sample means in Figure 6.4. Likewise, the distances between the sample means in Figure 6.5 and Figure 6.7 are equal. However, the distributions around the two means in Figure 6.6 are quite narrow, and the distributions around the two means in Figure 6.7 are rather broad. All other things being equal, sample size would be the most likely explanation of the apparent differences in dispersion. The sizes of the samples in Figure 6.6 are probably much larger than the sizes of the samples in Figure 6.7.

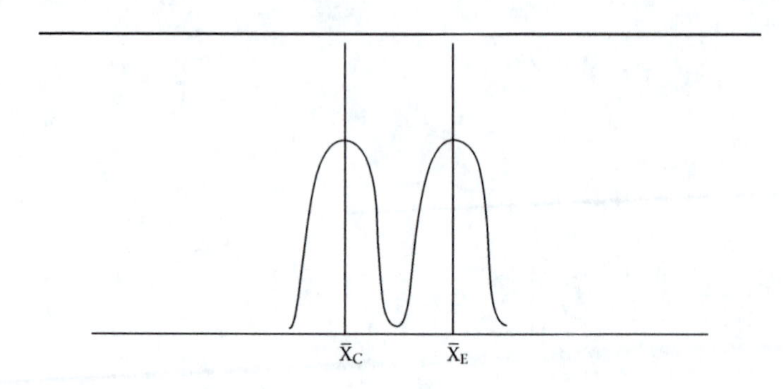

FIGURE 6.6
Small difference between sample means and large sample size.

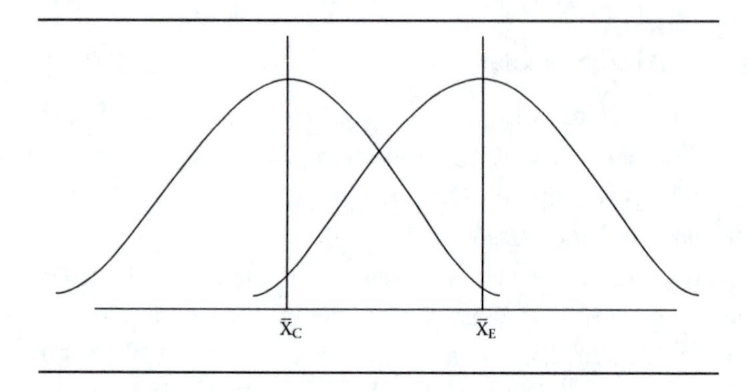

FIGURE 6.7
Large difference between sample means and small sample size.

Yes, I laid a trap with Figures 6.4 and 6.5. You are not alone if you concluded that the relatively close-together means in Figure 6.4 were much more likely to be from the same distribution than the far-apart means in Figure 6.5. Now, with dispersion added around each mean, you hopefully see the need to reevaluate the situation. These very simple figures show why statistics cannot make any inferences about the likelihood of sample differences without considering both central tendency and dispersion, with additional attention to the sizes of randomly drawn samples.

These four elements—mean, standard deviation (converted to standard error), sample size, and random selection of the samples—are the theoretical foundations of traditional statistical models that allow us to evaluate experimental data in order to see if something makes a difference. One

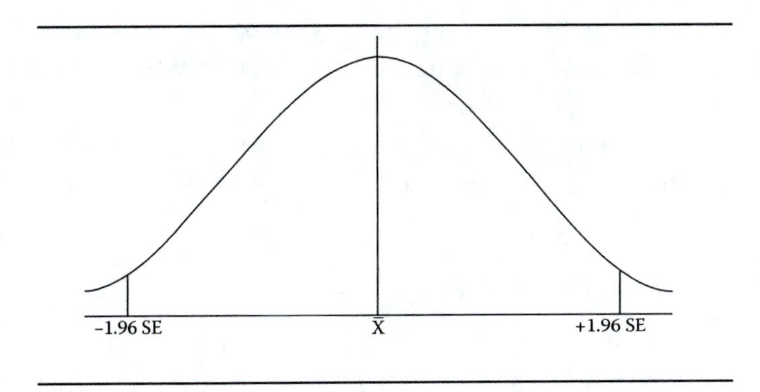

FIGURE 6.8
Distribution of population for evaluating sample differences.

more illustration will complete the picture of what happens conceptually when health professionals use statistical tools to transform data into information.

Figure 6.8 depicts a population that is normally distributed (that is, bell shaped), with the 95 percent confidence interval bounded by a point 1.96 standard errors below the mean and a point 1.96 standard errors above the mean. Conceptually, the equations of inferential statistics compare the difference between the means of two or more samples to the length of the 95 percent confidence interval on a scale expressed in units of standard deviation about the mean. This conversation is accomplished by the *z score,* which is explained at some length in most traditional statistics books. (Since you will never see the *z* score in the published report of a study, I see no reason to develop it for you here. Your success as a healthcare decision maker will not depend in any way on understanding this intermediate step in statistical computations.)

Accepting the Null Hypothesis

If the distance between the sample means (the numerator in the equations of inferential statistics) obtained through a proper experiment is short enough to fit within the 95 percent confidence interval of a population distribution with dispersion defined by the pooled variances of the samples (the denominator in the same equations), we can be 95 percent confident that the two samples came from the same population because we would draw sample means within this interval 95 times out of 100. In this case, the null hypothesis would be accepted, and the report of the study should conclude that the experimental effect—the new staffing model, the revised surgical procedure, the capitated payment mechanism, this book, or whatever else was being tested—did *not* make a statistically significant difference.

Rejecting the Null Hypothesis

Again assuming that all the theoretical conditions of science and statistics have been met, statistical analysis will reject the null hypothesis if the distance between the sample means is longer than the 95 percent confidence interval in the population distribution created by the statistical tool. (Of course, other confidence intervals can be defined thanks to the neat mathematical properties of a normal distribution. I have used a 95 percent confidence interval for illustrative purposes because it is so common, but 99 percent confidence intervals are also used.) Rejecting the null hypothesis means that the experimental effect *did* make a statistically significant difference at the chosen level of confidence.

An important qualification needs to be attached to this discussion of my hypothetical example. In accord with common practice, my null hypothesis states only that the experimental effect (reading this book) did not affect a health professional's understanding of statistics. Therefore, the null hypothesis could be rejected in one or two ways: *either* (1) using this book led to a significant improvement in statistical understanding as measured by the score on the final exam, which would be indicated by the mean of test scores from the experimental group being above the mean test score of the control group and beyond the 95 percent confidence interval, *or* (2) using this book led to a significant decrease in statistical understanding as measured by the test score, which would be indicated by the mean experimental group test score being below the comparable figure for the control group and likewise outside the confidence interval.

If the publisher and I wanted to bamboozle potential customers, our promotional materials could claim that a scientific study showed this book had a statistically significant impact on health professionals' understanding of statistics. This would be true, even if the individuals who used this book understood a whole lot *less* than those who did not use it. Potential buyers who did not know any better might be really impressed with the claim, even though the statistical test that supported our claim does not

STATISTICAL SIGNIFICANCE

Studies seem to take great pride in reporting that their findings are statistically significant. Exactly what does "statistically significant" mean? Is it a big deal? The answers suggest that statistical significance does not necessarily deserve the cosmic power commonly associated with it.

What does it mean? The rather tautological answer derives from the central limit theorem and the probabilistic properties of a normal distribution. By definition, we can expect to draw randomly a sample mean from within the 95 percent confidence interval 95 times out 100. Obviously, a mean outside this range will be drawn 5 times out of 100. The probability of drawing a mean outside the confidence interval is the *level of significance*, commonly called the **p** value.

Since the confidence interval and the level of significance account for all possible outcomes, probability theory says that the sum of their respective probabilities must equal 1.00 (100 percent). If the confidence interval equals 0.95 (95 percent), the **p** value must equal 0.05 (5 percent). Likewise, the 1 percent level of significance corresponds with a 99 percent confidence interval. In other words, *the level of significance is the probability that chance alone, not the experimental effect, accounts for the observed outcome* (such as the difference between the sample means in inferential statistics).

Is statistical significance a big deal? Not necessarily. The concept relates only to the null hypothesis, which states that the experimental effect makes zero difference. Rejecting the null hypothesis indicates that the experimental effect does make a difference—that is, that the difference is not zero—but it does not tell us how big the difference actually is. Indeed, *very small differences can be highly significant (that is, very unlikely to be explained by random sampling error, or chance) and very big differences can be explained by chance (for example, the result of a small sample size).*

Statistical significance by itself does not mean that a study has made a major discovery, nor does it necessarily endow a finding with practical significance. A savvy reader of research reports knows not to be unduly swayed by a low **p** value, recognizes the possibility of rejecting a true null hypothesis (Type I error) or accepting a false null hypothesis (Type II error) due to the vagaries of nature, and remembers that a single study proves nothing—no matter how statistically significant its finding.

distinguish between a desirable effect and an undesirable effect. Such is the potential problem with a *two-tailed test.*

When research has the potential to fall into this trap, the statistical analysis can be based on a *one-tailed test*, which fixes the decision point only at the upper or lower ends of the distribution, but not both. In other words, the analysis of the data from my hypothetical experiment would be conducted to reject the null hypothesis only when the mean of the experimental group is above the upper limit of the chosen confidence interval. The null hypothesis would also be restated to reflect the one-tailed approach: Reading *Statistical Analysis for Decision Makers in Healthcare* improves a health professional's understanding of statistics.

The single, upper-tailed approach would help me avoid the potential embarrassment of exposing an undesirable effect, but it also fails to forewarn potential buyers who undoubtedly would not want to buy the book if they knew it was likely to confuse them even more than *Statistics the Traditional Way.* One-tailed and two-tailed tests have their respective strengths and weaknesses. Be sure to look at the relationships of the sample means when you review a study that might affect your own decisions if the authors do not deal adequately with this issue. (Fortunately, most do.)

TEST STATISTICS

Recognizing that researchers would want to compare data from two or more samples in order to assess the probability that they were from the same population—which means accepting the null hypothesis that the experimental effect did not make a difference—statisticians have developed several statistical tools for this purpose. Each of the tools is really a mathematical equation where the difference between means is the numerator, and the standard error of the difference between means is the denominator. (Remember that the standard error in the denominator incorporates the dispersion and sample size of each compared distribution.)

Even though I do not believe you need to do the computations yourself in order to evaluate statistical information in your day-to-day roles as decision makers, I believe you do need to understand what happens when these equations are used to compute the values of the test statistics. One more point of theory needs to be explained before we look at the equations of the three basic tools of inferential statistics. To repeat an important point I made in the introduction, I do not care at all whether you can perform the computations, but I care a whole lot whether you can tell if someone else used the right statistical tool for the job. Computers can do the calculations, but only you can do the thinking!

Here is what happens with an inferential test statistic. The values of the descriptive statistics from each sample (mean, standard deviation, and sample size) are plugged into the numerator and denominator of the test statistic equation, which yields a quotient. The greater the value of the quotient, the greater the probability that the samples did not come from the same population. When the computed value of the test statistic passes a certain critical value (also called decision or threshold value), the distance between the means is effectively too long to fit within the chosen confidence interval on the measurement scale, the null hypothesis is rejected, and the study has demonstrated a statistically significant difference at the stated **p** value. On the other hand, if the computed value of the test statistic falls below the threshold value for the chosen level of significance, the null hypothesis is accepted because we are not sufficiently confident that the samples came from different populations.

The critical values for each of the test statistics are themselves computed with equations that reflect the probability characteristics of the underlying distribution for given sample sizes. These resulting values are usually presented for the most common levels of significance and sample sizes in tables at the back of computationally oriented statistics book. (A British brewmaster named Gossett developed the computations and the corresponding mathematical proof around the beginning of the twentieth century. He published them under the pseudonym Student, which is why the basic inferential test is called Student's *t*.)

For example, imagine that the computed value of a test statistic for significant differences is 3.26 with a sample size of 64, and the proverbial table at the back of the book gives a critical value of 2.17 at a 95 percent confidence interval (**p** value = 0.05) for this sample size. Since the computed value (3.26) from the study data is greater than the decision value (2.17) from the standard statistical table, the null hypothesis is rejected. We would conclude that the experimental effect made a difference greater than would be explained by sampling error only 5 times out of 100. Suppose we then decide to evaluate the result at a 99 percent confidence interval, for which the critical value is 3.47. We would accept the null hypothesis in this case because the computed value (3.26) is less than the value from the table (3.47), meaning that the difference found by the study is not statistically significant at **p** = 0.01.

In contrast to this example from the not-too-distant days when researchers had to go to a table at the back of the book to find the decision value of the test statistic, today's computer-based statistics programs simply print the **p** value at which the null hypothesis would be rejected, in other words, the value at which the results would be statistically significant. In the made-up example from the previous paragraph, the printout might simply say the null hypothesis would be rejected at **p** = 0.027.

This example shows that all sorts of games can be played with **p** values. Indeed, I have often suspected that some researchers were competing to see who could make a conclusion based on the smallest **p** value, as if a low level of significance were a sign of good research. Unsophisticated readers or researchers might jump quickly to the conclusion that the study found a really big difference when the results become statically significant at a very small **p** value, when all that really can be said, statistically speaking, is that the difference was very unlikely to be explained by chance.

t Test

The *t test* is the appropriate statistical tool to test for the possibility of significant difference in the independent means of two samples of parametric data. The basic mathematical formula for computing the value of the test statistic, *t*, is shown in Equation 6.2.

$$t = \frac{\overline{X}_i - \overline{X}_2}{\sqrt{\frac{S_1^{\,2}}{n_1} + \frac{S_2^{\,2}}{n_2}}} \tag{6.2}$$

In accord with the previous section's conceptual discussion, notice how the numerator of the basic *t*-test equation incorporates the difference between the means of the two samples and how the denominator pools the standard errors of the means to estimate the population distribution from which the sample means might have come. Statisticians have developed some variations in the *t*-test equation to reflect differences in the data, such as the possibility that the means are correlated—in other words, not independent. (Correlation is the subject of the next chapter, so do not worry if you do not understand this concept yet.)

F Test (Analysis of Variance)

The *t* test is the right tool for the statistical job when only two samples of parametric data need to be compared, but many studies require analyzing data from three or more groups. For example, I might be interested in comparing the statistical knowledge of three different groups: (1) students who used a traditional statistics text, (2) students who used this untraditional text, and (3) students who used no text at all but only attended lectures. This experiment would produce three different sets of sample data to test the null hypothesis that there is no difference in the distribution of postcourse test scores for students in each of the groups. (Accepting the null hypothesis would imply that using either book does not make a difference because students who only attended lectures are probably from the same population as those who also used a textbook. So that you and I will both feel better about the time we have each spent on this book, let us assume this null hypothesis is rejected.)

Statisticians have developed the *F* test for the purpose of analyzing parametric data from more than two samples. Also called analysis of variance and abbreviated ANOVA, the *F* test looks for significant differences in the ratio of two estimates of the potential variation within a population (see Equation 6.3).

$$F = \frac{S_b^{\,2}}{S_w^{\,2}} \tag{6.3}$$

Like the *t*-test equation, the formula for computing *F* is a ratio that incorporates the samples' means and dispersions. However, the *F* numerator is an estimate of the observed variability among the means of the three or more groups, which is not quite the same thing as the absolute scaled difference between two means used in the numerator of the *t* test. The numerator in the *F* ratio is computed with the means themselves, not the mean differences. The denominator is an estimate of the variability within groups, and the actual scores in the groups are used to compute it.

If the variances between the means are the same as the variances within the samples, the *F* ratio will equal unity (1.00). However, the value of the ratio will be greater than 1.00 if the sample means are relatively spread out while the variances within the sample distributions are relatively similar. Accepting or rejecting the null hypothesis for *F* is accomplished the same way the hypothesis is tested for *t*. The computed value for the study data is compared with the critical value from a statistical table, incorporating the degrees of freedom in the numerator and the denominator at the desired level of significance (**p** value). The null hypothesis is accepted if the value computed from the sample data is below the value from the standard table and rejected if the computed value is greater.

Since the purpose of this book is to help you understand concepts rather than struggle with computations, two final illustrations (Figure 6.9 and Figure 6.10) should help clarify how the *F* test works.

Although the means in both figures are the same (that is, the numerators of the corresponding *F* ratios will be the same), the distributions of sample data around the sample means obviously are not. The value of the

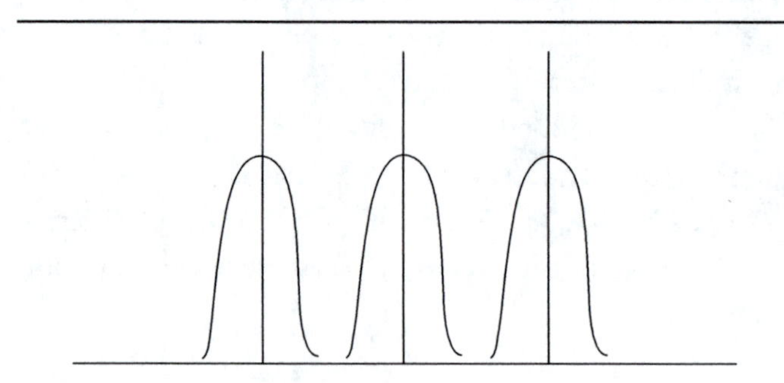

FIGURE 6.9
Small dispersion within sample distributions.

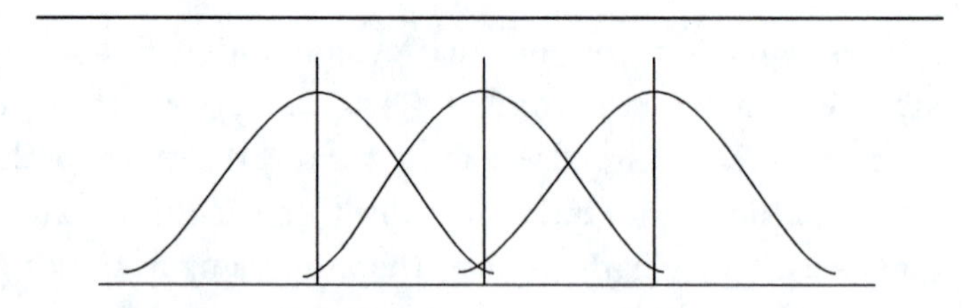

FIGURE 6.10
Large dispersion within sample distributions.

F ratio will be higher for the situation represented by Figure 6.9 because the variation in means (the numerator) is relatively large compared to the relatively small dispersions within the samples' distributions (the denominator). The denominator in the computed *F* ratio for Figure 6.10 will be larger due to the greater variation within the distributions, which yields a lower value for the ratio. (Just in case you have not thought much about fractions lately, the value of a ratio with a constant numerator goes down as the denominator goes up.) In other words, Figure 6.9 is more likely to reflect a statistically significant difference between the three samples; it has a higher *F*-ratio value.

Unfortunately, an *F* value that causes us to reject the null hypothesis does not indicate where the statistically significant difference actually is. It tells us that the overall differences in the samples are greater than the differences that would be explained by chance (that is, random sampling error), but it does not tell us which of the three groups accounts for the difference. Looking at the data may aid conjecture about the outcome, but statisticians disagree on what, if anything, can be done to identify specific factors that actually caused the statistically significant value of the *F* ratio. Exploring this issue is far beyond the scope of this book, but raising it is not. If you ever need to understand all the nuances of a study that draws conclusions on the basis of an *F* test, you should either discuss the issue with a statistician or read one of the many available books on analysis of variance.

You do need to know, however, that special *F* tests have been designed for a variety of situations. I have illustrated the general concept with a one-way analysis of variance, designated one-way because only one factor—the use of a book—varies in my hypothetical experiment. However, statisticians have customized *F* tests for experiments where several different possibilities are researched simultaneously. Imagine, for example, an experiment to test the combined impact of two different books and two different instructional techniques. This study would generate data for four samples: (1) students who used a traditional textbook in a statistics class, (2) students who used this untraditional book in a statistics class, (3) self-instructed students who used a traditional text, and (4) self-instructed students who used this untraditional text. A two-way analysis of variance would be used to test the null hypothesis in this case.

Fortunately, multidimensional *F* tests are rare in the types of studies conducted for the benefit of healthcare decision makers. Few who read this book will ever need to dig deeper into the issue. If you do encounter a complex analysis of variance in a study that might be important in your situation, you will also want to look very carefully at the research methodology because multidimensional studies are extremely difficult to conduct. Problems with control and sample size are tough to handle, once again reminding us that good statistics cannot compensate for bad research.

Chi-Square Test

The *t*-test statistic for comparing parametric data from two samples and the *F*-test statistic for comparing parametric data from three or more samples are extremely useful, but they are not the appropriate tools for all situations. Nonparametric (or categorical) data occur frequently and defensibly in healthcare studies, but they do not provide the sample-to-sample consistency in measurement that is central to the theory behind *t* and *F*.[3] Statisticians have developed another tool, *chi-square* (often indicated with the Greek X), for testing hypotheses when nonparametric data are involved.

Actually, chi-square is only one of many "distribution-free" test statistics tailored to the special characteristics of categorical data, but it is the most commonly used nonparametric tool. The theoretical foundations of chi-square, as explained in the following paragraphs, provide a good basis for understanding the general concept of nonparametric methods for testing hypotheses. However, just as different versions of *t* and *F* may be needed to reflect special situations involving parametric variables, one of the other nonparametric tests may be more appropriate than chi-square.

You probably do not need to become familiar with the many different nonparametric tests (such as Kruskal-Wallis one-way analysis of variance, Friedman two-way analysis of variance, Mann-Whitney U, Wilcoxon, Scheffé, and so forth) unless you decide to conduct your own studies and do your own statistical analysis. Understanding chi-square should be sufficient to make you an intelligent interpreter of the results of nonparametric tests used in other people's studies. In my experience, researchers who are smart enough to use nonparametric tests on categorical data are also usually smart enough to know which nonparametric test to use. Take

---

[3] Review the discussion in Chapter 3 if you need a refresher on the important difference between parametric and nonparametric measurements.

DOES TYPE OF DATA REALLY MATTER?

A theoretical difference of opinion divides statistical practitioners into two camps: those who are bothered by using parametric tests on nonparametric data (purists like me), and those who are not. Many researchers and authors of statistics books think we purists are too picky when we argue that the basic parametric tools, *t* and *F*, should not be used for the statistical analysis of nonparametric data.

The reason for the purist position is straightforward. As shown in the figures that illustrated the theoretical reasoning behind parametric tests (for example, Figure 6.6 and Figure 6.7), means and standard deviations from different samples of parametric data can be compared precisely because they are measured along a single scale with units of constant length and universal meaning. For example, a standard deviation of 2.50 years of age from an experimental group would be exactly twice as large as a standard deviation of 1.25 years of age from a control group.

If you are confused, look again at the figures in the middle of this chapter to see how parametric tests are made possible by consistently measured numbers with meaningful distributions. The *t* and *F* equations only make theoretical sense because the data from different samples can be added and subtracted. Taking the differences of means in the numerators and pooling dispersions in the denominators would be hard to justify if the samples were measured with dissimilar rulers—the proverbial problem of comparing apples and oranges.

By definition, nonparametric data do not have a consistent and meaningful measurement scale, and they are distribution free. For example, the distance between one value on a typical nonparametric scale (for example, "fair") and the value above it (for example, "good") is not necessarily equal to the distance to the value below it (for example, "poor"). So how can anyone justify analyzing nonparametric data with parametric tools, which were built on a theoretical assumption of consistent measurement and known distributions?

Researchers who are willing to analyze nonparametric data with parametric tests do not usually say the purists are wrong. They (the unpurists?) just believe a mismatch between type of statistical test and type of data does not matter very much. Purists believe it does. After all, why did statisticians take the time to develop nonparametric tests if parametric tools would do the job? Because the two types of data are so different, they should be worked with different tools!

The field of statistics does not have a supreme authority to resolve such disputes, so you have to decide this issue for yourself. I hope you will join the purist camp and be attentive to the proper match between type of data and statistical test.

a statistician to lunch if you feel the need to double-check the use and interpretation of statistics in a study that is especially important to you.

Since nonparametric data cannot be meaningfully summarized and compared via means and dispersions, another approach had to be developed to explore the possibility of statistically significant differences between two sets of sample data. The rather ingenious solution to the measurement problem is based not on descriptive statistics but on frequencies. Chi-square compares frequencies that were actually obtained with the frequencies that would be hypothetically obtained if there were no difference between them. The greater the deviation between the actual frequencies and the hypothetical frequencies if the groups were identical, the greater the likelihood that the samples are from different populations. Hence, the key to computing chi-square is the difference between obtained and expected frequencies.

To see how the process works, imagine a study to see if clinicians (such as nurses, physical therapists, and physicians) and nonclinicians (such as hospital administrators, clinic managers, and financial officers) understand statistics any differently after taking a one-semester statistics class. The subject data are purely categorical. The subjects are either clinicians or nonclinicians, observations that cannot be measured on any parametric scale. Under ideal conditions, we might be able to develop a parametric measure of statistical understanding, such as test scores on standardized examination, but reality and practicality will almost certainly dictate using a nonparametric measure of statistical understanding as well. Assume we can select (randomly, of course!) and survey 500 clinicians and 500 nonclinicians who have taken a one-semester course in statistics, and we determine whether they passed or failed the course.

TABLE 6.1

Chi-Square Analysis

	Pass	Fail
Clinicians	Observed – Expected = Difference	Observed – Expected = Difference
Nonclinicians	Observed – Expected = Difference	Observed – Expected = Difference

Conceptually, which means simplified for illustrative purposes, the process of analyzing categorical data with chi-square begins by putting the study data in a table of the form shown in Table 6.1. The chi-square statistic is computed by summing the squares of the differences in all four cells and dividing this sum by the expected value (in this case, the pass rate of all health professionals who take a one-semester statistics class). If the study (observed) data are identical to the expected rates of passing a statistics course, the value of chi-square will be zero. However, the value of chi-square will begin to increase as the study data in one or more of the cells begin to deviate from the expect value, and at some point the computed value of chi-square will become larger than a value that could be explained by chance (sampling error).

The critical or decision values for chi-square have been computed and are readily available in standard tables, so the final step in the process for hypothesis testing with chi-square is the same as the process used with *t* or *F*. The null hypothesis is accepted at the specified **p** value if the computed chi-square is below the decision value from a standard table. The null hypothesis is rejected if the computed value is greater than the decision value from the table, which means the samples are likely to be different populations at the specified level of significance.

Nonparametric test statistics such as chi-square are very useful when circumstances dictate the use of categorical data, but they are less precise than their parametric counterparts. For reasons I do not understand, many healthcare studies collect parametric data and then put them into categories before beginning data analysis. This practice is acceptable (as long as the categorized data are analyzed with nonparametric statistics!), but it is not necessary since computers can analyze the original parametric data just as fast as they can analyze the grouped data—and the results are more powerful in a statistical sense.

THE END OF THE TUNNEL

Congratulations, you have made it through the toughest part of statistics. You should now understand how descriptive statistics can be used to summarize large collections of data and how the summary numbers can be used to test hypotheses stating that data from different samples are from the same population. You have also seen how and why different statistical tools were developed to turn different types of data into useful information. I hope this approach has demystified statistics to the extent that you will know how statistics ought to work the next time you read a study.

By extension, I hope you have developed the knowledge-based courage to defend yourself against statistical malpractice. Indeed, if this approach has been really successful, you probably understand the theory of statistics better than most of the people who write the studies that fill the journals that clutter your desk. Do not be afraid to doubt what you read now that you know what is going on when statistical analysis takes place. Much of the healthcare literature we read is seriously flawed and deserves to be doubted. Making better decisions will be your long-term reward for being cautious when you review published studies. And for getting this far in the book, your immediate reward is the next chapter; it is very short and has lots of illustrations.

7

Relational Statistics: Studies of Relationships

Healthcare professionals have good reason to be interested in quantitative relationships. Their professional activities and their careers can be affected by answers to questions about the interactions between variables that may be under their control. For example, does average patient revenue decline as length of stay increases? Does cardiac risk increase with a patient's weight? Does a hospital administrator's income rise with experience? Is the mortality rate for a given procedure in a hospital inversely proportional to the number of times it is performed there each year? Does overall health status of a city's population decline as urban density increases? The decisions you make are likely to be affected by the results of studies that address questions such as these.

Statisticians have developed several mathematical tools for defining different dimensions of relationships. This chapter describes the basic concepts behind relational statistics and shows how they can help inform decision making in healthcare. It also identifies important limitations of relational studies so that you will not commit the errors of making unwarranted conclusions about causality or linearity. (Many people, especially journalists and politicians, regularly commit the error of drawing conclusions about causes from relational studies.)

DIRECTION OF RELATIONSHIPS

This general concept is really simple. Two variables can be related positively, negatively, or not at all. The specific statistical name for the relationship between two variables is *correlation*.

- *Positive correlation* exists when two variables move in the same direction at the same time. If one variable is increasing, the other variable tends to be increasing as well. Positive correlation also exists when they decline together because two negatives make a positive. Hence, the key to positive correlation is simultaneous movement in the same direction, not necessarily movement in a positive direction. Plausible examples of positively correlated variables in healthcare would include hospital executives' incomes and years of employment in a particular position, individual health status and years of education, physician visits and prescribed medications, and the number of journal articles read and clinical knowledge.

- *Negative correlation* exists when two correlated variables move in different directions. As one goes up, the other generally goes down (and vice versa). The key to negative correlation is the opposition in movement, not the existence of a negative direction in one of the variables. Indeed, by definition, one of the two negatively correlated variables has to be moving in a positive direction. As we saw in the previous paragraph, the correlation is positive if both variables decline at the same time. Examples of negative correlation from the world of healthcare could include a person's life expectancy and the number of cigarettes smoked, the average cost per procedure and the number of procedures performed (the economic principle of increasing marginal returns), an individual's weight and the average number of minutes per day spent exercising, and maybe even a health professional's understanding of statistics and the number of traditional statistics books read—on the theory that more reading only made matters worse.

- *Zero correlation* exists when two variables do not move together in any identifiable, repeated pattern. They are as likely to go in the same direction as in opposite directions. Nothing can be said about the expected nature of the relationship with any degree of confidence because the relationship appears to be random. Examples of uncorrelated variables might include the number of applicants to a nursing school and the amount of the application fee, physicians' class rank in medical school and annual income, or the number of primary care residents and the number of doctors selecting rural practical locations.

Studies that identify the direction of a relationship can certainly enhance a health professional's understanding of the environment in which she or he works, but knowing the direction of an association is not enough to lead to intelligent analysis and decision making. The strength of the relationship must also be known before any action is taken. Except in the case of zero correlation, the direction and the strength of a correlation are totally unrelated, so learn to look at both. (Above all, avoid the belief that positive relationships are more significant or stronger than negative relationships. I have encountered several students who had this misconception, presumably because positive numbers are greater than negative numbers. In statistical correlation, direction and magnitude must not be confused.)

STRENGTH OF RELATIONSHIPS

The statistical tool that measures the strength of relationships between two variables is the *coefficient of correlation*, most commonly abbreviated by the small letter *r*. Consistent with points made in the two previous paragraphs, the value of the coefficient of correlation is 0.00 when no association exists between two variables, but *r* will have a value different from 0.00 whenever a measurable relationship does exist. The highest possible value for the coefficient of correlation, *r*, is 1.00—which indicates a perfect correlation between two variables. Let us look at a few figures to see what this means in practical terms.

The figures that will be used to illustrate correlation are called *scatter plots*. Scatter plots are visual displays of the corresponding values of the variables that are being studied for correlation. All you need to create a scatter plot are the values of two variables—for example, variable *X* and variable *Y*—for all the cases in a sample or population. To show how this works, let us assume we have collected valid and reliable information on annual income and number of hours worked per year for a randomly selected sample of pharmacists. The data are presented in Table 7.1. (I am using a sample of only six cases to keep the example simple; random samples of 30 or more would be needed if we were actually doing a study and wanted to be reasonably certain that the sample represented the population.)

TABLE 7.1

Data from a Hypothetical Sample of Pharmacists

Pharmacists (Case #)	Variable *X* (Annual Income)	Variable *Y* (Hours Worked per Year)
1	$37,000	2,000
2	$45,000	1,900
3	$28,000	1,750
4	$49,000	2,200
5	$51,000	2,100
6	$54,000	1,900

Figure 7.1 shows the scatter plot created when these data are plotted against each other for the six pharmacists in the sample. The number next to each dot (data point) corresponds to the case number of the pharmacist.

A comment needs to be made about the common use of variables *X* and *Y* in relational statistics and graphic conventions you may remember from your high school algebra class. Do not confuse a scatter plot with the formal algebraic practice of assigning x to the independent variable and y to the dependent variable when graphing linear equations of the form $y = mx + b$. Assignment of variables to the x (horizontal) and y (vertical) axes is arbitrary in the process of creating scatter plots, but most computer programs for creating scatter plots will force you to use the conventional

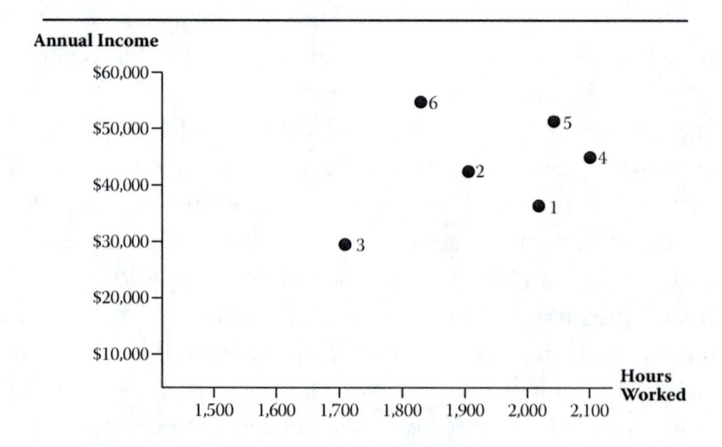

FIGURE 7.1

Scatter plot of hypothetical data from a sample of pharmacists.

X and *Y* framework with its implication of a functional (explanatory) relationship. As you will see later in this chapter, dependence (causality) is not addressed by the coefficient of correlation.

The three scatter plots presented in Figure 7.2 show different levels of positive correlation between variable *X* and variable *Y*. (Use your imagination to define the two variables in this illustration. I was tempted to define variable *X* as hours spent reading this book and variable *Y* as some measure of understanding statistics, but by now you should be able to create examples that are directly relevant to your own situation.) The direction is pretty obviously positive in all three illustrations because the data points, the joint values of the two variables for each case in the sample, have a perceptibly upward trend.

The strength of the relationships increases from left to right in the three illustrations in Figure 7.2. The correlation coefficients would be approximately $r = 0.20$ on the left, $r = 0.50$ in the middle, and $r = 0.80$ on the right. Notice in particular that the points begin to converge toward a straight line as the coefficient of correlation increases.

Figure 7.3 shows what happens when the relationship between data points is perfect. All the data points fall on a straight line, and the coefficient of correlation attains its limit value. In Figure 7.3, $r = +1.00$ since the two variables move in the same directions. The value would be $r = -1.00$ if the values of the two variables moved together with equal precision in the opposite direction. If the value of one variable is known when $r = 1.00$ or $r = -1.00$, the value of the other variable can be known with certainty—not estimated with error—because the relationship between the two variables is perfectly linear.

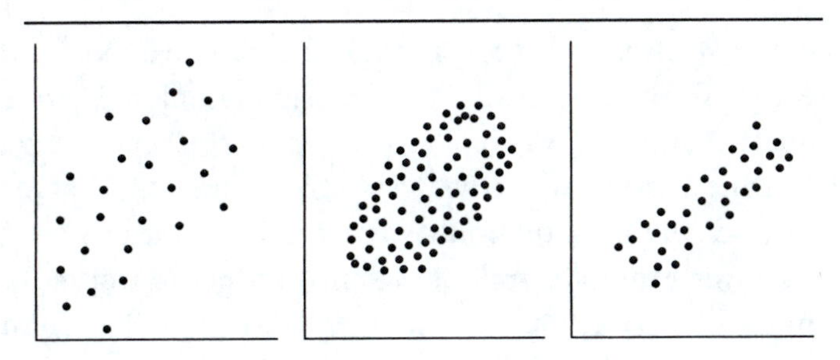

FIGURE 7.2
Scatter plots with increasingly positive correlation.

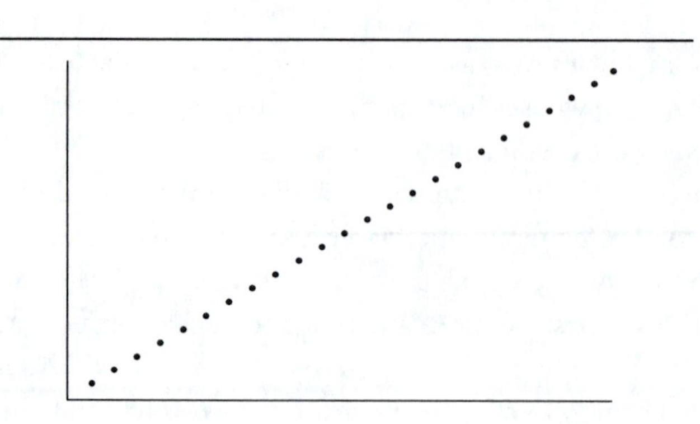

FIGURE 7.3
Scatter plot with perfect positive correlation.

TEST STATISTICS

Just as statisticians have developed many hypothesis-testing tools to reflect differences in data types and experimental circumstances, they have also developed a variety of test statistics for studying correlation. Using the right tool for the job is just as important in studying relationships as it is in testing for differences. Let us look at the basic parametric and nonparametric approaches to correlation analysis because they will handle most jobs as long as they are properly matched to the type of data. Out-of-the-ordinary situations will give you another excuse to invite a statistician to lunch for a discussion that goes beyond the scope of this text.

Parametric Correlation

The *Pearson product–moment coefficient* is the standard tool for examining the relationship between two parametric variables. (Karl Pearson, a British mathematician, was one of the founders of modern statistics in the late nineteenth and early twentieth centuries.) Pearson r was designed specifically for situations where the observations on X and Y for each case (for example, annual income and hours worked) are distributed normally, have approximately equal variability around their respective means, and have a linear relationship. Deviations from these assumptions correspondingly weaken the validity of Pearson r as a measure of correlation.

True to my promise to make this book conceptual and not mathematical, I will explain how Pearson *r* is computed but spare you the lengthy computations that are involved in the process. (If you really feel the need to see the math in all its detail, go to the chapter on correlation in your old statistics textbook. I guarantee you nothing has changed, no matter how long ago you took statistics.) See Equation 7.1 for Pearson *r*.

$$r = \frac{\sum xy}{NS_x S_y} \tag{7.1}$$

The numerator gives the sum of the products of the differences between each value of *X* and *Y* and the respective means of *X* and *Y* for each case in the sample. The denominator multiplies the product of the standard deviations for the distributions of *X* and *Y* by the sample size. If only one value of *Y* exists for each and every value of *X* and the distributions are symmetrical, the computed value of *r* will be one ($r = 1.00$ or -1.00) because the numerator and the denominator will be identical. On the other hand, if equally and proportionally varied values of the mean differences exist for both *X* and *Y*, the differences in the numerator will cancel one another, and the value of the numerator will be zero ($r = 0.00$).

Although I am pleased if the coefficient of correlation makes some sense following this simplified explanation, I do not believe that understanding the equation is essential for healthcare decision makers who will be interpreting the works of others rather than doing the work themselves. The two most important points to retain from this discussion are the importance of the assumptions behind the formula and the interpretation of the coefficient itself. In particular, do not base any decisions on Pearson *r* if you cannot be sure that the study's authors made sure it was the right tool for the job.

Nonparametric Correlation

One of the most obvious mistakes would be analyzing nonparametric data with Pearson *r*. The Pearson product–moment coefficient is based on very strict assumptions about the underlying distribution of the data, so it is not the right tool to use for categorical, distribution-free data. Fortunately, several statistics have been developed to examine correlation when one or both of the numbers are nonparametric. The *Spearman rank-difference coefficient of correlations*, often labeled ρ, is appropriate for estimating the relationship between two ranked variables. *Kendall's Tau* was developed for use when both variables are nonparametric.

Computer programs will generally provide a **p** value (level of significance) along with the correlation coefficient. The level of significance of the correlation coefficient—the probability that random sampling error explains the outcome—can also be a relevant factor if the analysis was conducted to estimate intervariable relationships within a population. Randomness and sample size are every bit as important in relational statistics as they were in inferential statistics when sample data are used to provide estimates of population parameters; the same conditions must be met. The sample size should contain at least 30 cases, and larger samples will reduce the role potentially played by chance. Also, the sample should have been drawn completely at random. The **p** value is meaningless when these conditions are violated.

LINEAR VERSUS NONLINEAR RELATIONSHIPS

The discussion so far has addressed only linear relationships: relationships that can be best expressed as straight lines over the entire relevant range of values of the two related variables. Indeed, the correlation coefficients covered in this chapter should not be used if the underlying relationship is anything other than linear. Figure 7.4 shows how a linear model can lead to erroneous and misleading conclusions when it is applied to nonlinear data.

Assuming that this scatter plot illustrates the paired relationships of two variables of parametric data, the approach taken by the Pearson product-moment correlation equation is conceptually represented by the straight line. Overall, the differences between the *X* and *Y* values and the line balance out because the number of points above and below the line is roughly equal.

However, the straight line does a lousy job of reflecting the actual distribution of points above and below the line. It fails to reflect the fact that almost all the points above the line are in the middle of the distribution

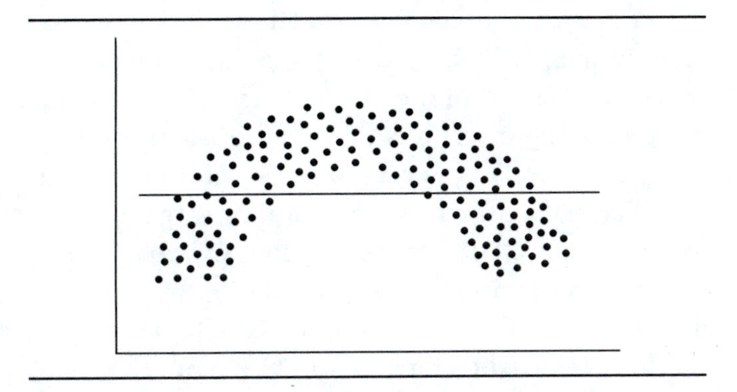

FIGURE 7.4
Nonlinear data and linear correlation line.

and all the points below the line are at the ends. (This situation lays the foundation for the subject of the next chapter, regression.)

Figure 7.5 presents two other nonlinear relationships that will be correspondingly misinterpreted if analyzed with the standard correlation coefficients presented in this chapter.

The illustration on the left obviously represents a cyclical relationship. The illustration on the right shows an exponentially increasing relationship. Again, both are characterized quite inaccurately when the correlation coefficient is computed with a model that assumes the data are all distributed in linear fashion (that is, symmetrically distributed on both sides of the line throughout the entire range).

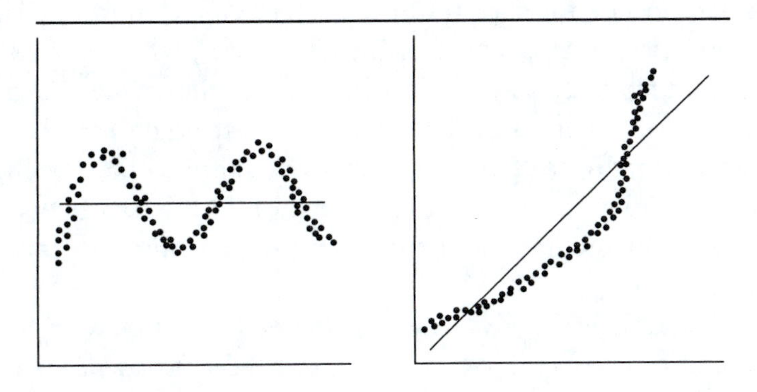

FIGURE 7.5
Nonlinear data sets.

The solution to each of the problems pictured in Figures 7.4 and 7.5 is to use a nonlinear model, one that matches the shape of the data with the appropriate mathematical function, such as a quadratic equation for Figure 7.4. The nonlinear world is the realm of statistical subspecialists and applied mathematicians. Expect to spend some extra time finding a properly qualified expert if you need help dealing with nonlinear data that are important to you, and do not even try to read books on the subject if you do not have skills in nonlinear math.

A good researcher will always look at the data before computing a correlation coefficient. Therefore, one of the signs of a good correlation study is evidence that the researchers examined scatter plots to ensure a proper match between the data and the statistical tool used to examine them. Be suspicious if a study reports correlations with any of the tools covered in this chapter without first verifying the linearity of the data. Indeed, analyzing nonlinear data with linear test statistics can be much more misleading than working nonparametric data with parametric tools.

CORRELATION AND CAUSALITY

The coefficient of correlation can help us understand important relationships and can even be used (with appropriate qualifications) as a predictive tool. However, anyone who uses the results of correlation studies must avoid the temptation to draw the conclusion that one of the variables is responsible for changes in the other. Correlation models only look at how two variables tend to move together, which is something very different from identifying the cause of the relative movements. For example, years of education and income are strongly correlated, but higher education by itself does not explain higher income. Intelligence, work habits, background, and interpersonal skills are factors that might actually explain income—and they may or may not be products of education.

By itself, correlation analysis does not embody control, *a priori* specification, and other scientific steps that are needed to address causality. However, a sophisticated application of the basic principles of correlation analysis can be used to study causality and explain outcomes. It is regression analysis, the topic of the next chapter.

8

Explanatory Statistics: Studies of Causality

As shown in the preceding chapter, the tools of relational statistics can be used to estimate the strength and direction of changes in two variables. But correlation coefficients do not tell us whether one of the variables causes the change in the other or whether the changes in both are due to the effect of some other variable not included in the analysis. We must, therefore, resist the temptation to draw conclusions about causality on the basis of relational statistics.

We need to work data with a different set of tools if we want to explain relationships in a way that allows predicting the new value of one variable as a result of changes in the values of other variables. *Regression analysis* is the statistical tool that allows us to explain and predict—that is, to make probabilistic statements about—causality. It allows us to study the simultaneous or lagged influence of several variables on another one.

Regression analysis is a relatively new tool, one of the few branches of statistics developed and popularized in the last half of the twentieth century. Indeed, the growth of explanatory statistics is itself explained by development of the computer because regression analysis requires hundreds or even thousands of calculations that are exceedingly tedious and time-consuming when performed by hand. Without computers, explanatory statistics would not be as popular as it is today. (Do not worry; I am not about to break my no-math promise by working through a sample problem and losing you this late in the book!)

STATISTICAL MODELS

The difference between inferential and explanatory statistics is a direct reflection of the difference between a hypothesis and a model. The first step in inferential statistics is to specify a hypothesis, but the first step in regression analysis is to construct a theoretical model. As shown in Chapter 6, a null hypothesis is a verbal statement that an experimental effect does not make a difference. Testing the hypothesis requires an experiment. The null hypothesis is rejected or accepted at a specified level of significance.

On the other hand, a model is an equation that expresses the theoretical mathematical relationships between two or more variables. A model does not require an experiment. Rather, it requires collecting data—past values of all the variables in the model's equation—and solving the model's equation with these values. The result is measured by how well the data fit the theoretical equation. A model is good or bad, not accepted or rejected. Model builders keep respecifying the equation by adding and deleting variables, either until they are happy with the fit or until they conclude that a good model is not possible under the circumstances.

Building mathematical models is fun. You may wish to try your hand at it someday. You do not have to go to the expense, in terms of time and money, of conducting an experiment. All you need is an accessible database that includes observations on the variables that need to be included in the model. However, all the requirements of good data are still critical. Models are no better than the validity and reliability of the data used to construct them. Randomness and sample size are also relevant if the model uses data from a sample to draw conclusions that will be extended to a population. All of this book's previous comments on the principles of science and the quality of data still hold. Good model building cannot compensate for bad science or bad data. "Garbage in, garbage out" is as relevant to statistical models as it is to scientific experiments. Therefore, the quality of data and data-collection methods must be one of your primary concerns when you review studies that use the tools of explanatory statistics.

Data-related problems are rather abundant in the published literature that uses regression analysis. Do not be surprised if you encounter some seriously flawed models, and do not rely on their conclusions when you make your own on-the-job decisions. On the other hand, do not prejudge all such studies; each has to be evaluated on its own merits. Some good work is being done with this relatively new tool. I hope that the overall quality of model-based studies will improve as researchers and journal editors become more familiar with the correct use of the method.

Regression models have been developed for both linear and nonlinear applications, so researchers must be careful to match the model and the data accordingly. As was the case with relational statistics, the results obtained with a linear regression model are only meaningful if the analyzed data are related in linear fashion. A nonlinear model must be used if the data lie along a curve rather than a straight line.

Nonlinear regression analysis is extremely complicated; it goes well beyond the scope of introductory statistics books, including this one. Consequently, the discussion in this chapter pertains only to linear regression analysis. If you want to learn about nonlinear models, you will have to consult a book dedicated specifically to them—but you would probably be wasting your time because nonlinear regression is hardly ever used in the literature prepared for healthcare decision makers.

Linear regression, on the other hand, is used relatively often for studies in the healthcare literature. Hence, one of your first tasks in evaluating a published study that employs regression analysis is to make sure the report's authors verified the linearity of the data before they used the linear model. In effect, the regression model in most common use today, known as the ordinary least squares (OLS) model, calculates the slope of a straight line that minimizes the distances between each data point and the line. The reported results are suspect if you have any reason to believe that linear regression analysis was used to analyze nonlinear data.

One other introductory point needs to be made before we look at the regression model and its underlying assumptions. Regression is conceptually similar to correlation, but regression has one major advantage. While the correlation analysis can address the interactions of only two variables at one time, several variables can be studied simultaneously with a regression equation. Therefore, the specific focus of the following discussion of explanatory statistics is OLS multivariate linear regression.

THE BASIC EQUATION OF MULTIVARIATE LINEAR REGRESSION

The basic model of multivariate linear regression analysis is built around an equation of the following form:

$$Y = \beta_1 X_1 + \beta_2 X_2 + \beta_3 X_3 + \ldots + \beta_n X_n + e \quad (8.1)$$

Each of the letters in this equation has a specific meaning and its own name:

- The letter to the left of the equal sign, Y, represents the *dependent variable*, the factor whose value is explained by the other variables in the model. A model can only have one dependent variable.
- The X factors to the right of the equal sign represent the *independent variables* that explain the value of Y. The β (beta) in front of each X is the coefficient that actually quantifies the impact of the corresponding independent variable on the value of the dependent variable (Y). The dots between the two plus signs indicate a continuing series of n variables (e.g., if the model had five independent variables, the series would extend to $\beta_5 X_5$).
- The values of each β are the solutions to the *regression equation*, called the regression coefficients. They are the unknowns determined by the computer when it performs all those complex calculations I mentioned a few paragraphs back.
- The e at the end of the equation is the *error term*; it is also called the alpha in some statistics books. Either way (e or α), this is the constant that balances the equation, roughly equivalent to the remainder in a division problem.

To illustrate the interpretation and application of a multivariate linear regression equation with a simple example, let us pretend that a study of the number of children in two-parent families (the dependent variable, Y) had been studied as a function of the number of years the parents had been married (independent variable X_1) and net family income (independent variable X_2). Suppose we collect the observations on all three variables from 200 randomly selected two-parent families, enter the values into our computer, and tell the computer to solve the regression equation. The answer on the computer printout should look like something like this:

$$Y = 0.11\,X_1 + 0.026\,X_2 - 0.27$$

Once our computer has used the sample of 200 randomly selected families to calculate the beta regression coefficients (+0.11 and +0.026) and the error term (−0.27), we can use the model to predict the number of children we would expect to find in any other two-parent family from the same population. All we need to do is multiply the number of years the parents have been married (X_1) by 0.11, add the family's net income (X_2) multiplied by 0.026, and then subtract 0.27. For example, if we know the parents of another family have been married for 14 years and earn a net family income of $57,800 per year, the model estimates the number of offspring as 0.11(14) + 0.026 (57,800) – 0.27, or 3.31 children.

If a regression model has good predictive value according to statistical criteria described in the following pages, the newly selected couple is likely to have three children. A proper regression model can be very helpful to decision makers; it allows them to predict the unknown value of important dependent variables based on the values of independent variables for which measurements are available.

SPECIFYING THE MODEL

Selecting the best independent variables to include in a regression equation, a process known as specifying the model, is arguably the most important task performed by researchers who use explanatory statistics in their studies. Specification is the skill that separates good statistical modelers from bad ones. I have developed an example to show you how the task ought to be performed. Some of the essential theoretical assumptions behind linear regression are embodied in proper specification and are also covered in this discussion.

A priori (that is, before the fact) thinking is as important to explanatory statistics and predicting as it is to inferential statistics and hypothesis testing because the human mind can spot random trends in the numbers. Consequently, a study based on regression analysis should include evidence that the authors specified the model before looking at the data. Hats off to the study where data were collected after the researchers decided which variables to include in the equation. Be suspicious of the study where the researchers specified the model after reviewing the data.

To begin the proper *a priori* process of selecting variables, researchers should precisely define the model's dependent variable (Y). I have selected the physician-to-population ratio in urban areas as Y for this example because healthcare decision makers can benefit from understanding how the supply of physicians is determined in their market. I am limiting this illustration to urban areas since we have ample evidence that the supply of physicians in rural areas is determined by a different set of factors. Being very specific at this stage is important because a model will be weakened if it covers too much territory (literally or figuratively).

The next step is to identify factors that might explain how many physicians will be located in an urban area. This exercise will create a list of independent variables that can be considered for inclusion in the regression analysis. A brainstorming session involving people who know something about physicians' locational decisions might produce the following list:

- Population
- Per capita income
- Unemployment rate
- Number of hospital beds
- Combined tax rate (sales + property + income).
- Percent of population with health insurance coverage
- Number of cultural facilities (museums, concert halls, botanic gardens, etc.)
- Distance to major recreational areas (mountains, lakes or ocean, etc.)
- Number of college and professional sports teams
- Per-student spending on public education

Some fundamental issues must be addressed before the final list of independent variables is selected for analysis. One of the most important concerns is the *independence of the independent variables.* The multivariate linear regression model is based on an assumption that the independent variables are just that—independent and uncorrelated. The results of regression analysis are distorted to the extent that any of the independent (*X*) variables are correlated with each other or with the error term, so good researchers will carefully consider the possibility of relationships between the explanatory factors before specifying the model.

The potentially correlated independent (explanatory) variables need to be identified and discussed. Ideally, the best one will be retained for the model, and the others will be discarded. In my example, I would probably have to choose between per capita income and per-student spending on public education because they have a good possibility of being related. (Personally, I would retain per-student spending on education because the area's income is related to other factors as well. Besides, based on my experience as an economist, I believe income figures are neither valid nor reliable.) Population and the number of hospital beds are also likely to be correlated. This situation might be resolved by combining both into a single variable: hospital beds per 1,000 residents.

The next issue for discussion is the distribution of values of the independent variables. For regression analysis to work properly, the variances (dispersions) within the distributions of each independent variable should be constant. This condition is *homoscedasticity*. Consequently, an *a priori* discussion of the independent variables should consider whether the distributions may have different shapes. For example, if the distribution of unemployment rates in the sample cities is bimodal and skewed while the values of all other variables are distributed normally, unemployment should not be included in the equation.

Each potential violation of the assumptions of the multivariate linear regression model has a name:

- *Multicollinearity* occurs when two or more of the independent (*X*) variables are correlated.

- *Autocorrelation* results if the independent variables are correlated with the error term.

- *Heteroscedasticity* (see previous paragraph) exists when the data violate the assumption of constant variances in the distributions of values for the independent variables.

The occurrence of any of these problems can reduce the precision of the regression estimates. Of course, the fundamental assumption of linear relationships between the dependent variable and each of the independent variables must also be met in order for the model to have an acceptable degree of explanatory power.

The number of independent variables to include in the model needs to be determined after the initial list has been reviewed and revised. Regression equations can have too many variables, a condition known as *overspecification*. Basically, the complexity of the computational process causes a model's power to decline as the number of variables increases. Statisticians differ in their opinions about the maximum number of variables to include in a regression equation.

My graduate school statistics professors contended that five or six independent variables was a good limit, and my subsequent experience has made me comfortable with this maximum. Indeed, including more than half a dozen variables is almost always a good sign that the researchers did not take time to think before running the data. You have good reason to be suspicious of a study if its results are based on an overspecified regression model. (I have been amazed to see equations with 10 to 20 variables in prestigious journals in the recent past. The results of such studies should be labeled with a user's advisory warning!)

In my opinion, you also have good reason to be suspicious of a study if the researchers used a procedure known as *stepwise regression* to select the variables for the equation. The stepwise approach lets the computer find the "best fit" by adding independent variables one at a time in different combinations until the model reaches its maximum explanatory power. A stepwise model involves no *a priori* thinking by the researchers, and that bothers me a lot because the stepwise protocol will find random associations and give them the appearance of being causal.

EVALUATING A MULTIVARIATE LINEAR REGRESSION MODEL

The model can be run once the regression equation is specified with a clearly defined dependent variable and an appropriate number of independent variables. To explain how the process would proceed from there, let us assume that we have methodically selected the following variables for our effort to identify factors that explain the doctor supply:

Y = Number of physicians in a metropolitan statistical area (MSA)
X_1 = Hospital beds per 1,000 residents
X_2 = Percent of population with health insurance coverage
X_3 = Number of cultural facilities

I would probably include two or three more independent variables if I were really going to test this model; three will suffice for illustrative purposes. Assume also that I randomly select 60 metropolitan statistical areas (MSAs) from a list of the hundreds of MSAs in the United States in order to keep the entire process manageable. Now we need numbers to crunch, so we would look for databases that provide valid and reliable information about the selected variables. After a visit to a reference library and a few online searches of digital databases, I should be able to gather the data I need: the same-year observations on all four variables (Y, X_1, X_2, and X_3) for the randomly selected 60 cities in the sample. The 240 data points are entered into the on-screen format of a statistics program that will calculate the unknowns, the three regression coefficients, and the error term.

Once the calculations are completed, the computer printout will give us a result that should look something like the equation below.

$$Y = 4.03\ X_1 - 0.07\ X_2 + 41.92\ X_3 + 2.07$$
$$(2.24) \qquad (3.71) \qquad (0.47)$$

$$F = 6.35\ (p = 0.09)$$

$$R^2 = 0.32$$

This same information should be included in the published report of any study that uses linear regression analysis to explain a causal relationship, so learning the meaning of these test statistics on the printout will help you decide whether to rely on researchers' claims.

I must forewarn you, however, that most authors do not present the regression test statistics in their articles. When it comes to providing the information you and I need in order to decide whether to have confidence in a regression equation, the prevailing attitude in healthcare journals seems to be "Trust us" or "You wouldn't understand if we told you." Well, I am absolutely not inclined to trust authors' interpretations of test statistics I cannot see for myself—especially when I know that many researchers do not understand the finer points of regression analysis. (There ought to be a law preventing anyone from using regression software until he or she has passed a test on the basic model.) I urge you to be equally cautious. Do not base your own decisions on a regression-based study that fails to present the test statistics.

The first test statistic in a typical printout is the *t* value, which appears in parentheses below each regression coefficient. It tests the null hypothesis that the computed coefficient for each independent variable is not significantly different from zero in its paired relationship with the dependent variable. If the *t* value is above the critical value, the relationship is statistically significant at the desired level of significance. If the *t* value is below the threshold value, chance sampling error could account for the observed relationship. As a rule of thumb, $t = 2.00$ is the decision point for $\mathbf{p} = 0.05$, so you do not need to consult a table if you are comfortable operating with a 95 percent confidence interval.

To see how this works in practice, look at the *t* value for independent variable X_1. It is 2.24, which is above 2.00 (the threshold value at $p = 0.05$), so the 4.03 multiplier computed for the number of hospital beds per 1,000 residents is statistically significant. The coefficient for X_2 is also significant, but note that the value of the multiplier (0.07) is very small. In other words, the percentage of the population with insurance coverage has an effect greater than that which would be expected by chance, but its effect is almost inconsequential. On the other hand, the coefficient for X_3 is quite large (41.92), but it is not statistically significant. The coefficient's value could be explained by sampling error.

I have intentionally invented computed coefficients (β and *t*-test values) to focus your attention on the difference between magnitude of effect and statistical significance. The two phenomena are different. A big difference can be explained by sampling error, and a small difference can be very real (that is, not explained by the luck of the draw). Make sure that a study clarifies this distinction before you base your own decisions on its reported results.

The *sign* of each regression coefficient, positive (+) or negative (–), is also important to evaluating the independent variable's effect on the dependent variable. The sign does not matter if the *t* value suggests that a coefficient's true value is not statistically significant from zero. In this situation, the independent variable does not explain changes in the dependent variable, which means the *X* is probably not a cause of changes in *Y*.

On the other hand, a coefficient's sign does matter in a statistical sense if the computed *t* value suggests that the relationship between the *Y* and an *X* is significantly different from zero. If the sign is positive, *X* and *Y* move in the same direction. An increase in the *X* term causes an increase in *Y*, and a decrease in *X* causes a decrease in *Y*. (Do not forget, however, that the actual magnitude of the changes might be very small.) When the sign of the computed coefficient is negative, a change in an *X* leads to the opposite change in the *Y*. The value of the dependent variable decreases as the value of the independent variable increases, and vice versa.

Looking at the signs of significant independent variables is one of the most useful features of multivariate linear regression analysis. Decision makers can benefit from knowing not only that a change in one variable probably causes a measurable change in another but also that the change occurs in a predictable direction. This information can be very useful to a healthcare executive or clinician who controls independent variables, as long as he or she knows that the regression model was properly crafted.

Following *t* values and signs for the regression coefficients, the next statistics presented in a typical regression printout is usually the *F value*. Whereas the *t* values are related only to the paired relationships between the dependent variable and each independent variable (such as *Y* and X_1, *Y* and X_2, *Y* and X_3, and so on), the *F* value encompasses the entire model. As was the case with hypothesis testing in inferential statistics, the critical value for *F* must be taken from a standard table that takes into account sample size and degrees of freedom. If the computed *F* is greater than the standard table value, the relationships in the model are likely to be real at the selected confidence interval. If not, the observed relationships between the dependent and independent variables could theoretically be explained by the luck of the draw (sampling error). Hence, you can see why knowing the *F* value is an important factor in your own evaluation of a published study. Authors who withhold the information may have something to hide, or they may not know what they are doing.

The last statistic presented in a typical regression printout is the R^2 *value*, technically called the coefficient of multiple determination. *R-squared* (R^2) *is the percent of variation in the dependent variable explained by the independent variables included in the model.* For example, if $R^2 = 0.32$ for my illustrative model, 32 percent of the variation in physician supply in the same data from 200 cities is explained by the number of hospital beds, the relative extent of health insurance coverage, and the number of cultural facilities. I strongly suggest that you also learn to think in terms of the converse: $R^2 = 0.32$ means that 68 percent of the variation in the number of physicians is *not* explained by the model's independent variables.

With this defense mechanism, you will not be misled by the many studies that try to make a big deal of a small *R*-squared. How big does R^2 need to be? That is a question you have to answer for yourself, but I do not take a model seriously until it explains at least 60 percent of the variation in the dependent variable; that is, the value of R^2 would have to be at least 0.60 before I would make a decision based on a regression analysis that otherwise met all the tests of believability.

I would not give any credence whatsoever to a study that fails to report a model's *R*-squared value, even if some *t* and *F* values are significant. Just as a statistically significant variable can have a very small effect on the dependent variable, several statistically significant variables can explain very little variation in the overall model. In my experience, the value of R^2 is not presented in at least half of all healthcare studies that use regression analysis, which makes them just as useless as a research report that claimed statistically significant differences without report the **p** value. (Journal editors, take note: Require full statistical disclosure as a condition of publishing any study.)

In an ideal world, the report of a regression-based study would also include the values of the test statistic for multicollinearity, heteroscedasticity, and autocorrelation. These three tests show respectively how well the model conforms with the theoretical assumptions of independence between the *X* variables, equal variances within the distributions, and zero relationship between the independent variables and the error term. However, these three statistics are almost never reported, not even in the most rigorous journals, so I will not present them in this illustrative discussion. As long as authors demonstrate appropriate attention to *t*, *F*, and *R*-squared, I am generally willing to give them the benefit of the doubt on the tests of compliance with model assumptions. However, I would contact the researchers for a discussion if a model were especially important to my situation.

So far, this chapter has presented the principal linear regression model developed for analyzing parametric data. Indeed, the model's underlying theoretical requirement of equal variances in the distributions can only be met by parametric data. By definition, categorical data violate this assumption because they are distribution free. The results are correspondingly flawed when OLS multivariate regression is used to compute coefficients for models that involve nonparametric data. The practice of using a parametric model to explain relationships involving nonparametric data is extremely common, but that does not make it right. I trust you will not base important decisions on studies that do it.

The good news is that regression models have been developed to deal with nonparametric data. Under the general heading of logistic regression analysis, these models (for example, Logit, Probit) use probability theory to create assumed distributions for the values of categorical variables. The nonparametric models provide useful tools for analyzing distribution-free data, and they deserve respect when they are used on their own terms. However, they are not as powerful as parametric models applied to parametric data.

Once again, I am accustomed to being called a purist for taking the position that nonparametric data should not be analyzed with parametric explanatory models. Researchers who are not bothered by this practice generally argue that a mismatched analysis is better than no analysis at all, especially when categorical data are the only available numbers. (I still believe strongly that no study at all is better than a bad study.) We purists are not swayed by their position because we can usually think of valid, parametric ways to measure the subject of interest.

For example, many nonparametric variables are created by lumping parametric data into categories, such as taking the income data from individual tax forms and reconstructing it in groups such as under $20,000, $20,000–$29,999, $30,000–$39,999, and so on. The simple solution to this problem is to return to the original, pregrouped data and use them for the analysis. Another misapplication is the common practice of arbitrarily assigning a number to sex (for example, 0 = female, 1 = male) and including sex as an independent variable in a regression model. This problem can be easily and appropriately solved by specifying two different equations, one for female and one for male. Indeed, this may be the only appropriate approach in light of growing knowledge about clinically significant differences between men and women.

REFINING REGRESSION MODELS

Assuming that all steps in the regression analysis have been properly followed, a big question usually remains after the first run of the data: Can the model be improved? There would seem to be plenty of room for improvement if the t and F values show little or no statistical significance and R^2 shows that very little variation in the dependent variable is explained by the independent variables. In many such instances, the model can be strengthened by respecification that begins with rethinking the relationships and replacing some or all of the variables. This process is legitimate as long as it involves careful thought prior to each respecification. It becomes very questionable once it deteriorates to trying every possible combination of variables to see if anything works, the human equivalent of stepwise regression.

Many researchers seem to feel they have failed if their repeated efforts at respecification do not produce statistically significant relationships and good explanatory power. Consider another possibility—namely, that no clear and consistent order is there to be found. A regression model may fail to identify strong explanatory relationships for the pure and simple fact that none exists. Chaotic systems do not have any causal relationships to reveal, and many aspects of today's healthcare system arguably are in a state of chaos.

We would all be better served if journals would print properly conducted explanatory studies that found nothing significant. In my opinion, this would be preferable to the current practice of publishing only those studies that "find" significant causal relationships but use poor methodology. A well-executed regression model that reveals nothing may be very useful to today's healthcare decision maker. On the other hand, a flawed study that reports erroneous conclusions can make matters worse.

APPLICATIONS OF EXPLANATORY STATISTICS

Regression analysis has a lot to offer when it is properly applied. This chapter's hypothetical study of factors that might explain the physician supply is a simplified example of work actually done with the tool by many researchers (myself included) in the 1970s and 1980s. Explanatory models have a lot to offer in situations where experiments are infeasible but historical data are readily available—as long as the data are good. There are several promising areas for further applications of explanatory statistics:

- *Quality of care*, particularly studies of relationships between outcomes and clinical factors such as providers' experience and education, frequency of performing a procedure, institutional size, and so on.
- *Financial performance*, including studies to analyze profit or loss, economies of scale, and input or labor substitution.
- *Demand for health services*, with special emphasis on patients' characteristics that explain visits to providers, severity of medical conditions, and use of ancillary services.
- *Enrollment in payment plans*, examining potentially causal factors such as health status, health history, and income.

A good explanatory model in areas such as these can produce valuable information to support administrative and clinical decisions or public policy. For example, pretend you are vice president of marketing for a growing capitation plan. Regression analysis of enrollment data for the past 20 quarters suggests that the number of hours spent in direct sales and spending for online ads are both statistically significant variables in explaining new memberships in the plan, but the relationship between enrollment and sales effort is positive, and the relationship with online ads is negative. You do not have to be a financial genius to see how such information can help you allocate next year's sales budget between direct marketing and Web sites.

Explanatory studies can generally help you identify variables worthy of your attention. Knowing the magnitude of statistically significant relationships helps determine where effort and resources can best be expended, and knowing the direction of these relationships helps focus on which explanatory factors to promote and which to impede. Expect to see even more use of explanatory statistics, but be sure to apply the lessons of this chapter to make sure the studies are worthy of your consideration.

Postscript: Statistics in Perspective

I have decided to end this book the way it began, with some comments about my particular perspective on statistics and your use of the information contained in these pages. The book was written to give you the skills needed to be a competent evaluator and user of statistical analysis performed by other people. If I have been successful in meeting my goal for this book, you may actually now know more about the underlying theory and proper practice of statistics than many people who conduct the studies and write the reports that cross your desk. Others may know how to compute a particular test statistic, but you know whether it should have been computed in the first place.

Here are a few parting thoughts that will help you get the maximum benefit out of the time and effort you have spent reading this book and traditional texts on the subject:

1. My "purist" view of the theory and practice of statistics is not a sign of disrespect for quantitative analysis in general or statistics in particular. Statistical tools are not perfect, but they are the best tools we have until the next generation of analytical techniques is perfected and introduced into daily practice. Statistical analysis most definitely can provide us with useful information and knowledge when its tools are used correctly. My critical comments about sloppy or inappropriate use of statistics absolutely must not be interpreted as cynical criticism of statistics. (On the other hand, these comments are critical of sloppy researchers.) We can use statistical tools well as long as we are aware of their limitations.

2. The next generation of analytical techniques has been under development for the past decade, and its first tools are beginning to have an impressive impact on the analysis of information in healthcare. Computer networks and advanced software are beginning to liberate us from analysis limited by the capabilities of a slide rule. For

example, new analytical tools with amazing graphical interfaces allow us to observe changes continuously over time; we do not need to make inferences because we can see what is actually happening under different (i.e., experimental) conditions. Nonlinear analysis is also becoming much more accessible. I am most excited by the evolution of bioinformatics, a rapidly growing set of tools that allows researchers to understand biological activity at the molecular level—providing a superior alternative to randomized controlled trials in many applications. I urge you to familiarize yourself with the advancements in quantitative analysis by reading relevant articles in *Science*, *Nature*, *Scientific American*, the *New York Times*, and other publications that report on the process (not just the results) of research.

3. Pay close attention to quality of data. Expect good numbers from others. Make sure that measurement and data storage are handled properly whenever they are under your control. The next generation of analytical techniques will be largely worthless if the quality of numbers is not improved in the process. If I am cynical, even hostile, toward any issue addressed in this book, it is the prevailing insensitivity to validity and reliability of data. If I have one favor to ask of you, it is to join me in making sure that we have good numbers to analyze and in exposing bad numbers in studies that are intended to induce changes in healthcare.

4. Remember that *statistics is only a tool to help you make decisions*. A well-done study provides useful information, but the human mind alone has the capacity to turn that information into knowledge. Test statistics impart no information about the political and personal contexts within which healthcare decision makers must operate today and in the foreseeable future. I have faith that good thinkers will always be worth more than good statisticians—especially when good thinking is accompanied with creative vision about the exciting realm of possibilities for bringing healthcare into the twenty-first century.

Good luck to you as you try to change healthcare for the better. I hope this book provides some help for dealing successfully with the many challenges that lie ahead.

Index

"f" indicates material in figures.
"n" indicates material in footnotes.
"t" indicates material in tables.

D

E

F

G

H

I

N

O

P

Q

R

About the Author

Jeffrey C. Bauer, Ph.D., is a Chicago-based partner in healthcare management consulting and director of the healthcare futures practice for Affiliated Computer Services, Inc. (ACS). With more than 18 years of experience as a consultant, he assists hospitals and other provider organizations with strategic planning and visioning, leadership education, technology assessment, and service line transformation. Dr. Bauer also spent 17 years as a professor at two medical schools and served as health policy adviser for the governor of Colorado. A nationally respected medical economist and health futurist, Dr. Bauer has published 165 articles, books, and videos. He speaks frequently to national audiences about key trends in healthcare, medical science, technology, reimbursement, information systems, public policy, health reform, and creative problem-solving. In his numerous publications and presentations, he forecasts the future of healthcare and describes practical, creative approaches to improving the delivery system. Dr. Bauer is quoted often in the national press and writes regularly for professional journals that cover the business of healthcare. He was a Ford Foundation independent scholar, a Fulbright scholar (Switzerland), and a Kellogg Foundation National Fellow. He holds a B.A. from Colorado College and a Ph.D. in economics from the University of Colorado at Boulder.